MAN
ON ANOTHER WORLD

MAN

ON ANOTHER WORLD

GÖSTA EHRENSVÄRD

TRANSLATED BY LENNART AND KAJSA RODÉN

THE UNIVERSITY OF CHICAGO PRESS / Chicago and London

Original edition: *Expansion: Liv i universum.* © 1961 by Gösta Ehrensvärd
Published by Bokförlaget Aldus/Bonniers, Stockholm, 1961

Library of Congress Catalog Card Number: 65-17287

THE UNIVERSITY OF CHICAGO PRESS, CHICAGO & LONDON
The University of Toronto Press, Toronto 5, Canada

 Published 1965. Composed and printed by THE UNIVERSITY OF CHICAGO PRESS, *Chicago, Illinois, U.S.A.*

PREFACE

The purpose of this book is to suggest some common factors of life as a universal phenomenon, especially the expansive expression of life that involves a certain degree of awareness. Thus, it is not concerned with the development that is represented, from our local point of view, by the evolutionary processes of the past five hundred million years but rather with something seen from a wider perspective—the complex developmental pattern of conscious life, which is here characterized as "expansion."

The concept of expansion is interpreted from different viewpoints. In part we shall be concerned with the general prerequisites for the evolution of endogenous life, not only on the earth, but in various places in the universe, a complicated expression of the general evolutionary tendency of possible chemical reactions. Certain biochemical lines of thought at this level link up with an earlier survey, *Life: Origin and Development*, which has appeared in this series. Expansion also signifies the natural tendency of forms of life that have developed a certain degree of consciousness to explore their closest surroundings, first of all the closest planetary neighborhood. This activity of man has been treated as an illustration of something that is universally possible, without any great profoundness, but only as an attempt to carry the discussion toward a more advanced level than that of our present activities within our home grounds. The presentation touches on a number of problems that are difficult to approach concerning the progress of the evolution of awareness toward advanced consciousness.

Biochemical evolution, the development of conscious life, local expansion into space as a whole represent the preliminary stages of a broadening of the sphere of interest, an expansion of thought, into which technical progress enters as a catalytic activation process within the development of thinking as such. From this point of view this book, in all its primitiveness, is an attempt to suggest a tendency of evolution, no more—a suggestion concerning the technical possibilities of reaching, through contact with the surrounding cosmos, a certain understanding of the concept of life, unconscious or conscious, in all its expressions, and through this eventually also an understanding of conscious life itself. A future contact with some distant part of the cosmos where, for some reason, thought and reflection have already reached an advanced level would certainly be a stimulating surprise. On our planet, we obviously have a lot to be done in that respect.

CONTENTS

PART I. HOME REGION

I. THOUGHTS BEFORE THE DEPARTURE / 3

II. POINT OF DEPARTURE / 10

III. FROM THE PERSPECTIVE OF THE MOON / 30

IV. PREPARATIONS FOR AN ENCOUNTER / 47

V. VENUS: THE VEILED UNKNOWN / 65

VI. MARS: THE PLANET WITH POSSIBILITIES / 84

VII. OUTPOSTS / 104

VIII. THE RETURN HOME / 115

PART II. TOWARD THE LIMITS

IX. SIGNALS FROM SPACE / 123

X. ADVANCED CONSCIOUSNESS / 137

XI. THE LONG-TERM AIM / 154

XII. THE TASK / 167

COMMENT / 173

INDEX / 175

MAN
ON ANOTHER WORLD

PART I. HOME REGION

CHAPTER I

THOUGHTS BEFORE THE DEPARTURE

There is a feeling of departure in the air. We feel it more and more every year. It began about 1945 with the first indications that something new was being prepared. We saw the effects on a large and frightening scale in the V-2 and atomic fission; both were results of a new technical development, the first trial stages of which became synonymous with devastation and mass death. And we found that both ideas (guided missiles and large-scale nuclear fission) were useful—first, of course, to combine in the way which is easiest for us primitive men: long-distance weapons with nuclear warheads. And 1946 came with peace and the cold war and a climate suitable for a detailed study of the technical potentialities of these ideas for practical use. Simultaneously came the possibility of a new discovery which would extend the boundaries of traditional thought. This new discovery could give us new and not immediately destructive ways in which to express our aggression and anxiety; it could give us the conquest of new areas, journeys toward new goals. It could lead to expansion into space.

And 1948, 1949, and 1950 came, and we saw the smoke streaks of the rockets as they flew over White Sands in New Mexico and over unknown bases in the Aral area. Nuclear physics became a science in its own right, and radioactive isotopes became an everyday reality both in laboratories and in practical life. In 1951, 1952, 1953, and 1954, the reports about conditions in the outer atmospheric strata of our planet became more numerous and the use of atomic fission for peaceful energy production in the next few years became a realizable goal. And in 1955 and 1956 the first atomic energy power plants were connected to the distribution networks in England and the United States. The first satellites started to orbit the earth in 1957 and 1958. Somewhat blunted by the speed of these developments, we witnessed the target shooting toward the moon in 1959 and 1960 with polite interest. Later, in 1963, came the transmission of signals from the Venus fly-by probe and in 1964 the first close-up pictures of the moon's surface. Now we are in 1965.

It may be wrong to say that we are blunted; that is probably not the right word. In general, we human beings are attracted to what is new; some of us are attracted by the unknown. Those participating in the research on manned spaceships, which have become a reality this year, have the natural enthusiasm for the job and for its realization in practice that is characteristic of intelligent professional workers. Plans go ahead at full speed both in the East and in the West, and the rivalry over the final goal of the first phase of development, with all its implications of national and political prestige—to land one of the species of Homo sapiens on the moon—gives the work a special stimulus. As for the rest of us, on the periphery of the technical preparations, our general attitude toward the whole undertaking is marked by our interest in the headlines, which we follow with the unqualified admiration due to highly skilful work. In addition there is also a certain fascination with the possibilities of adventure and perhaps also some anxiety over what the consequences might be. Perhaps, too, there is a general feeling of unpreparedness. As technically trained beings we may feel prepared, but as individuals or as a society we do not.

As individual human beings, we are beginning to realize that through the increasing multiplication of the possibilities of science the field is becoming too large to survey as a whole or even in a limited section. The time has passed when, as in the eighteenth century, a man with a newly awakened enthusiasm could share in the latest developments in all the fields of scientific research, that naïvely charming time when man thought that knowledge as such would become a unifying factor between human beings. The nineteenth century came, and science became the sciences, first as a small number of limited fields such as chemistry, physics, geology, and biology, but later divided into more and more specialized areas. Universal knowledge was succeeded by specialized knowledge, and the development of this tendency up to our times is too well known to merit further comment. One reason for the anxiety we feel may be that the highly developed professional knowledge which is necessary nowadays to solve more and more complicated problems has led us to the position in which an individual feels alien to any development as a whole. It is not only that the technical vocabularies of specialized fields are developing into a number of different dialects, comprehensible only to an initiated few and creating language barriers between professional people working in even very closely related fields, but there is also something else.

We are now in a phase in which individual efforts in the scientific field become parts of larger projects. This development in itself is a natural and practical method of organization of the specialists who are collaborating in an attempt to solve problems of a high degree of difficulty; this happens in all fields of human activity and is by no means new in principle. But what we now are beginning to notice is the distressing result: these projects become so large that they catalyze their own continued existence and development in a certain impersonal way. They start to function as systems of such a high degree of organization that they seem somehow to be organisms. The activity of the project as a whole can no longer be surveyed from the viewpoint of the participating units, or hardly even from the viewpoint of those who are considered to be in control of the participating units.

Earlier in history there have been projects both of a local and of a general nature in which innumerable units have worked and died for aims which have been within the compass of one idea. The large-scale construction of pyramids, the crusades, the conquest of continents, war in general, and recently a very well-organized example, the creation of totalitarian states with their subordination of the individual to the dictates of the leaders—these are all such projects. In earlier periods, however, the aims were simple and well defined, and the immediate results of the project would have been either quite clear or surveyable without any great effort. In our own time of 1965 it is more difficult to do this, and one factor makes it especially so. The surveyors of the past, that is, those individuals with an acknowledged position as such and with a capacity for independent long-range thinking, are now to be found in the units of well-co-ordinated group activities. They are employed as top-level specialists, but their chances of comprehending any project in its entirety are decreasing. They cannot understand what the immediate consequences of their projects will be for mankind in general, and they have virtually no chance of being able to lay down rational long-range directions.

In our time this is true of projects of the most varying character: the purely humanitarian, such as the campaigns against cancer, malaria, and mental illness; the support of the underdeveloped areas of the world; social experimental projects, more or less well prepared, but with the possibility of further development; and finally, as a natural and inevitable consequence of our inherent primitivity, well-organized projects which aim to defend or spread a special social regional development even at the cost of a brutal neglect of more humanitarian considerations.

In the development and activity of all these projects we can see the many conflicts which arise when two different organizations with different aims come into close contact. Within each organization there is always a large number of specialists. With regard to basic knowledge, they are at the same level, but even though they may represent the same sector, they have different orders. They work for the project wholeheartedly, but they work for different reasons. They are partly driven by the forces within the organization; they are partly driven by their own interests which demand the solution of the immediate tasks which have been assigned to them; and they are driven by a concern for their families. They have hardly any time left over in which to consider the more general problems of the practical application of their project and its consequences. This is true for us all, both for those who have uncomplicated special tasks in a complicated world and for those who, as a result of circumstances, have been selected to initiate, organize, or control projects of a wider nature.

This state of affairs should not necessarily give rise to doubts about the adequacy of man's insight and farsightedness. In many ways we should be grateful for the technical evolution which has given primitive man many external comforts. We are disturbed every now and then, it is true, by the obvious mistakes and misunderstandings over the actual distribution of the comforts, but after all we have been given a multitude of interesting tasks with which to occupy ourselves. We have to face the fact that we are still such simple beings that most of what we achieve in our small world is emergency solutions, the urgency of which leaves us little time in which to consider the complicated results. That the technical development which we have originated is now getting the upper hand over its creators is something that we must realize, not in an indifferent or resigned manner, but with the self-irony that gives perspective and the desire to grow with the task. From this viewpoint we see that the years of exploration of space and of our neighboring planets that lie ahead are projects in which we all participate in one way or another without realizing the consequences to which this project will lead us: a large-scale experiment initiated by a primitive, naïve, true-hearted being, lacking the ability to plan more than a few years ahead and to understand more than the immediate goals but still somewhat aware of the possibility of further development.

It is possible that this experiment, begun by man in an unprepared state, will in the long run give impetus to an evolution in areas which up to now have been notably neglected. Besides all the new technical experience—which will undoubtedly be gained through great sacrifices both of individuals and of material—the first journeys toward the boundaries of our planetary system and the first colonization in space must lead to a consideration of wider problems.

First, it is clear that the pioneers sent out to these distant places through great economic efforts must be well educated in a general sense; they must have a high degree of specialization in one or more fields; and above all they must be extremely well adjusted in order to deal with highly varying conditions. They must be able to stand—with composure—both long periods of isolation in a formidably uncomfortable outpost and long periods of contact with their fellow men in conditions of extreme overcrowdedness. Under these conditions man's mind will be tried hard, and it might be that his experiences will, in the long run, change the attitudes of those who stay at home toward upbringing, education, and social behavioral patterns. Furthermore, it is possible that the daily contact with problems outside our own local earthly ones will eventually leave its traces on our attitude toward existence as such. We may discover the proper relationship between man and the surrounding universe and—perhaps as a consequence—between human beings themselves. Our first contact with life on one of our neighboring planets will surely provoke us to think about our own life.

This is the background against which we have to view our present explorations. We have no reason to be proud of ourselves either because of the basic knowledge or the technical skill which we have developed over the past hundred years, even if it does enable us to inspect for the first time the nearest regions in our planetary system. Sooner or later such a project would have been undertaken anyway, and we are well aware of the fact that at the present time the real driving force behind the realization of interplanetary journeys is of a purely political and military, not an idealistic, nature. This is an inevitable result of the automatic ways in which the organizations concerned go about realizing their ambitions.

The greatest mistake we can now make in the planning of our expansion is to assume that this project has never been undertaken before in the universe and to use it or other similar competing projects for propaganda purposes to prove how good we are from a limited or local point of view. We must become aware of the fact that every civilization, every conscious form of life, that has gained a certain insight into science (it has taken us some two hundred years) wants or feels obliged to extend its individuality through space colonization. It is inevitable that problems about life, its origin and its evolution, will become actualized through this process. Above all, we may, within a foreseeable time, establish contact with life in the planetary neighborhood, which will force us to redefine our concept of life on the basis that life is a universal phenomenon.

If we are clear about all this, there is much of great interest to be discovered during the next few years by those engaged in one or another phase of the project of expansion. There will be discoveries which will clarify from a universal point of view our own local evolution. There will be the chance that we will gradually become more sensitive—by our experiences in our own planetary system—to signals from space which may inform us that our own particular worries are not unique in the universe.

We may now realize how much our intellectual equipment leaves to be desired, that the world is ruled by a wisdom that is remarkably poor; but what a strange power the primitive lust for adventure has over our senses! Those responsible for projects both now and in the future will undoubtedly declare that nothing is left to chance, that everything is calculated in advance. For all projects. But the fact remains that certain individuals will experience something completely new in unknown regions and will have an opportunity to manage momentary situations of an unsuspected degree of difficulty without the meddlesome interference of the central authorities. Men will experience great elemental pleasure over their discoveries in the cosmic desolateness, and these personal experiences will be evaluated after their return from the first venture into space. We are still primitive beings learning the hard way, learning to see things both as a whole and in detail, learning more, living more, so that we can gradually achieve longer and longer excursions toward our limits. We all feel tense and apprehensive—more than ever before at this very moment. Our chance has come too suddenly. We are unprepared. We do not know what we will encounter and what will happen in the long run. But it is time to depart. There will be a great deal to tell when we meet again.

CHAPTER II

POINT OF DEPARTURE

The technical background for the present (1965) exploration of our closest planets is a combination of well-known visual and photographic methods of observation in the classic astronomic tradition and new methods of space probes, indicating, at this preliminary stage, promising possibilities. In the first group we can include, of course, those instruments which are now part of the standard equipment of the larger observatories all around the world, the giant 200-inch reflector of the Mount Palomar Institute being the largest of its kind. These large reflectors are of primary importance for planet observation, but they have many other functions to perform. There are also those instruments which have been developed during the last few years for radioastronomic purposes, with the 200-foot mirror at the Jodrell Bank observatory in England as the leading construction for short-wave reception. A still larger (600-foot) reflector is being built in Puerto Rico, which, when completed, will be the world's most powerful radiotelescope. (See Fig. 1.)

These instruments together with additional equipment for spectrum analysis cover a wave-length range from ultraviolet to long-wave infrared, and in the short-wave range, 1 to 100 cm. As a result, observations of our neighboring planets from very different aspects can be made with these instruments. An analysis of present observation data is given in Tables 1A and 1B.

From our observations a general picture of our planet system has been gradually built up. General data about the shapes of orbits, mass conditions, and rotation elements are given in tables in various parts of this book (see pp. 32, 66, 85, 106). Figure 2 serves as an illustration of the distances and dimensions involved.

It is clear, however, from the compilations of astronomical data about our planet system that there are obvious discrepancies in the accuracy of our knowledge about the local surface conditions on the planets compared with our knowledge about orbit shapes,

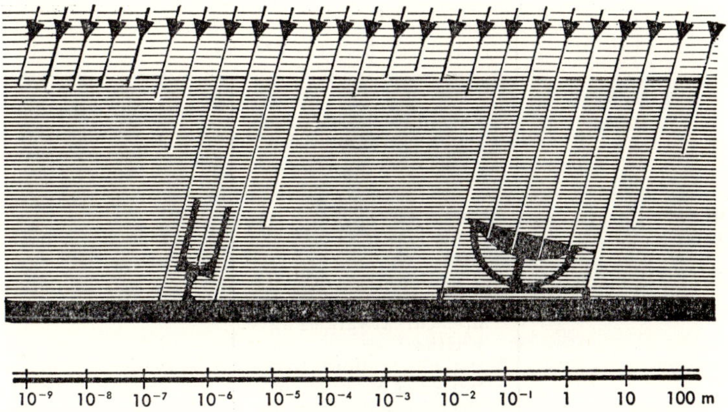

FIG. 1.—Diagrammatic representation of the way in which electromagnetic radiation penetrates our atmosphere and ionosphere. The narrow wave-length band between 4×10^{-7} and 8×10^{-7} m, which penetrates relatively easily the atmospheric nitrogen, oxygen, carbon dioxide, water vapor, and other components, represents what we call "light." Outside this band there is an ultraviolet radiation region, between 2×10^{-7} and 4×10^{-7} m, and a region of long-wave infrared, between 8×10^{-7} and 2×10^{-6} m. Outside this band width of 2×10^{-7} to 2×10^{-6} m, the components of the atmosphere absorb the incoming radiation almost completely with the exception of the special fraction that passes the "window" between 3×10^{-3} and 3 m. This radiation, which represents what we call ultrashort radio waves, can penetrate the charged, ionized stratum of the atmosphere that we call the ionosphere. Longer wave lengths—from 10 to 10,000 m—are reflected against the ionosphere. Thus our "eyes" by which we can see the universe are, on the one hand, optical instruments such as telescopes and, on the other hand, large reflectors which are built so that they can catch the ultrashort radio waves.

masses, and distances. Even with planets which we can observe without the disturbance of a thick atmosphere—especially the moon, Mercury, and Mars—our observations from a distance have been able to tell us little about topographic and geologic conditions. Knowledge about these conditions would give us a standard of comparison by which we could judge the history of our own planet as well as the development of local life on our earth. Table 1B, which sets out what we think we know about atmospheric and

TABLE 1A

Planetary Data	Methods	
	Visual and Photographic Registration in the Ultraviolet-Visible-Infrared Region	Radiotechnical Registration in the Short-Wave and Ultrashort-Wave Range
Orbital elements	Visual observations	
Distance	Calculations from orbital data	Direct measurements by radar
Size, mass, diameter	Calculations from orbital data. Micrometric measurements at various wave lengths	
Rotation	Visual observations. In certain cases observations of the change in frequency of spectral lines. Doppler effect	Observations of changes in frequency in the short-wave range. Radar data
Surface temperature	Direct thermoelectric measurements of emission from the outer atmosphere, in certain cases from the ground surface	Direct observations from the ground surface of the target region
Atmosphere, structure	Visual observations in various wave-length ranges	Registration of atmospheric disturbances in the short-wave signals from certain planets
Composition	Spectroscopic analysis, notably absorption spectra	Registration of the absorption of certain cosmic short-wave noise in the atmosphere of certain planets
Planetary surface, structure	Visual observations in various wave-length ranges	Radar observations
Composition	Spectral analysis of reflected light	

TABLE 1B

Object	Period of Rotation	Planetary Surface Conditions	Atmosphere	Temperature (° C)
Mercury	Probably the same as the period of revolution —88 days	Visual contrasts	None?	+200° to +400°
Venus	Unknown; between 15–60 days	Unknown	Dense, lacking contours. Carbon dioxide dominates	+15° to +55° in the upper atmosphere. >+200° at the ground surface
Mars	24 hr, 37 min, 22.6 sec	Visual contrasts, notably the formation of polar caps. Observable variations	Thin, with cloud formations. Small amounts of carbon dioxide and water	−60° to +15° depending on the day-night conditions and the seasons
Asteroids	Unknown	Unknown	None	Approx. −100° to +10°, depending on the orbit
Jupiter	9 hr, 50 min to 9 hr, 55 min	Unknown	Extremely dense; turbulent. Methane and ammonia, hydrogen and helium	−120°
Saturn	10 hr, 14 min to 10 hr, 38 min	Unknown (possibly lacking a defined surface)	Same as atmosphere of Jupiter	Approx. −150°
Uranus	10 hr, 45 min	Unknown	Same as atmosphere of Jupiter	Approx. −180°
Neptune	16 hr	Unknown	Same as atmosphere of Jupiter	Approx. −200°
Pluto	Unknown	Unknown	None?	Approx. −220°
The moon	27 days, 7 hr, 43 min	Mapped for 2/5 of the surface	None?	−120° to +120°

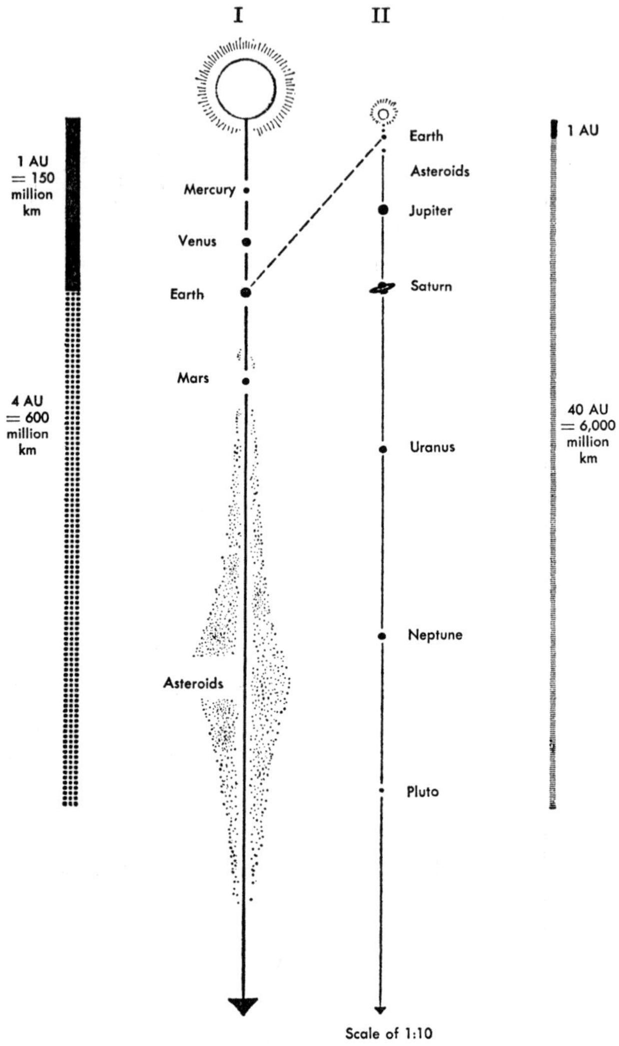

FIG. 2.—Schematic representation of the distances within our planetary system. The relative distances from the sun to Mercury, Venus, earth, Mars, and the asteroid fields outside the Mars orbit are shown on the left (I). The unit of measurement used is the average distance between the earth and the sun—one astronomic unit (AU). On the right (II) there is a similar diagram on a reduced scale of 1:10 showing the distances of other planets from the sun.

ground conditions on our neighboring planets, illustrates very clearly the deficiencies in our astronomic knowledge.

It is at this point that modern space techniques come into the picture. We are becoming more and more aware that our observations from a distance will never be able to give us more than a limited amount of information; that in some ways they have definite limitations even if further developments in techniques were to take place. Questions such as whether the craters on the moon are the results of earlier volcanic activity or of the impact of meteorites and the classic question of the possibility of endogenous life on Mars or Venus can hardly be resolved by observations from a distance but only by on-the-spot observations.

The current developments in space travel as a means for astronomic research fall into two categories: unmanned remotely controlled space probes and manned spaceships capable of making close observations as well as landing and making the return journey. Both projects are closely interrelated and developments in one will catalyze the other and vice versa.

The technical background for the construction of rockets and satellites—both for local use and for interplanetary expeditions—has become so complicated in the last few years that it is almost impossible to give a summary of the general technology and the construction data in this field. Such a summary would largely be outside the scope of this volume. From the mass of data we extract only a few general points to discuss here.

Let us start with the project of robot satellites and unmanned data collectors in interplanetary traffic. The largest of the former type are already able to carry from about one to three tons of equipment, including instruments for receiving orders, registering and transmitting data, and power units. The present-day development of compact and miniature instruments and power-sources will probably eliminate the need for long-distance probes to weigh more than one ton even for the solution of relatively complicated tasks. Problems which ten years ago were considered insoluble have already been solved by quite simple methods.

The Russian Lunik III, only 270 kg in useful weight, is an example of this. One week after its launching on October 4, 1959, it had already supplied data from a previously unknown astronomic region: the hidden surface of moon. According to the statement issued about it, the satellite was spherical; it rotated freely throughout its journey along an extremely elliptical orbit, the starting direction of which represented a mastery of celestial calculation. At the correct moment, just before its passage behind the moon's disk, the satellite reacted to a signal which released small gas jets; these stopped the satellite's rotation and made it possible for photographs of the moon's surface to be taken. The following series of photographs was taken during a period of forty minutes by two 35 mm cameras with 50 and 20 cm focal distances respectively. Another signal then released a new set of gas jets which rotated the satellite again, thus securing in the interior of the rocket the even temperature which is essential for the automatic development of the film. On its return journey at approximately 40,000 km from the earth, a new signal brought an electronic device into operation which scanned the developed film and telemetered it to earth, where a reconstruction of what had been observed at a distance of 460,000 km could be made. That the quality of these pictures was not very high is not surprising, for after all this was the first attempt to obtain knowledge from a hitherto unknown cosmic area. That the simple methods which were employed did work in practice was, however, of fundamental importance.

This was 1959, and many new things in the field of space technology have happened since then. In particular, the technique for long-distance telemetry has advanced far beyond our expectations. The system used for the Russian Lunik III project—return of the space probe to the vicinity of the earth for low-power telemetry of photographic data—was indeed ingeniously simple, but has now been superseded by more efficient methods. First of all, the power sources of space probes have become more advanced in the last five years. The first primitive solar batteries, utilizing the heat of solar radiation upon a series of thermoelements, have been much improved, enabling an output of about 20 w/m^2 from the semiconductor units placed around the central body of the space probe. Other sources of energy are being explored. The SNAP project in the United States has two lines of approach. One is to utilize the heat of disintegrating radioactive isotopes—strontium-90, polonium-210, and others—for electricity production via thermoelectric energy conversion. Already in practical use are 60 w units, but the aim is to construct units with an output of several kilowatts. The more advanced types of the SNAP series are planned to have an output of more than 100 kw, with the electricity production based on uranium fission and energy conversion by turboelectric units of very compact construction. It is estimated that the ratio weight/kilowatt should in the future approach 100 kg/kw, maybe still less.

To all these advances in the construction of compact power units can be added the new developments in telemetry involving *directed beams* for communication. It is now possible to equip a space probe with a parabolic reflector transmitter continuously directed toward home base with the aid of a photoelectric system using a certain star as reference point. The success of the United States Venus fly-by mission, 1963–64, was due to this principle in combination with modern improvements of solar batteries. What effect will this truly amazing development in the field of long-distance telemetry have on the present United States space mission to Mars (1964–65)? In the meantime, waiting for the signals, let us examine briefly the principles of telemetry.

In data transmission or telemetry, every piece of information transmitted needs a certain pulse amplitude and a certain pulse frequency per unit of time. Both have simple relationship with the available transmitter energy: the higher the pulse amplitude, the fewer the number of pulses per interval. A remote transmitter, which has to transmit with a high pulse amplitude in order to be heard at all, can transmit, with a limited amount of available energy, a limited number of impulses—perhaps only a few—in one interval, whereas a close transmitter can transmit a multitude of pulse series of low amplitude. Information transmitted directly from a remote transmitter will therefore of necessity be scanty in comparison with the detailed information which a returning satellite, even with a transmitter of a few watts capacity, can transmit when it is nearer the earth.

Multiplicity of detail is essential in the exploration of the surface of a planet. Information transmitted by a satellite, for example, on pressure, ionization, or meteorite strikes, or on photo data or color impressions, etc.—perhaps twenty different kinds of information in all—is now generally recorded as code numbers, every piece of information being categorized in a serial order according to a predetermined scheme. These series of signs are then sorted by the electronic computers of the receiver and stacked up according to the transmitting organ. Thus it is possible to record in an extremely short time all the data about a satellite at any given moment.

This will result in the future—perhaps in a few years—in an accumulation of a series of recordings about Mars and Venus obtained by phototelevision techniques from a distance of some tens of thousand of km from the surface of the planet in question. The question is, however, what will these data yield for those who are interested in something more than an example of human ingenuity or in outline mapping? Would a space photograph of our own earth, taken from an altitude of 20,000 m, indicate the existence of life? (We disregard the possibility that the well-lighted areas of our large cities might be visible at night.) What does a photograph of the moon tell us, even when it is highly detailed?

Close-up photographs from moving satellites will probably predate much more highly complex methods of observation: *soft landings* followed by the transmission of information from a transmitter in direct contact with the ground surface or the atmosphere. In principle two different methods of observation can be imagined:

1. The satellite is directed toward the planet, braked, and directed into an orbit around the planet; it then lands and procures information both by registering data and by collecting material such as ground and atmospheric samples. By remote control it is then started on its return journey to earth to deliver its collected data, both material and registered.

2. The satellite is directed toward the planet as in the first method; it lands and procures data, which it then telemeters to earth. The data, including automatic analyses of ground material and atmospheric components as well as special analyses of possible organic material, are examined on the spot.

The first of these two alternatives will present problems similar to those of sending one or two observers to the planet. The weight of the machinery plus the fuel necessary for a return trip from Venus or Mars is, at least as far as Venus is concerned, theoretically as great as that of a multistage rocket for launching a smaller satellite into space. The ratio of the starting weight to the essential useful weight, which at present is about 500 to 1, could probably in the future be brought down to 200 to 1, but hardly any lower. This means that a projectile of five tons must be landed on Venus in order to return a 25 kg projectile to the earth. Let us suppose that everything works according to plan. The information needed, including both materials and registered data, could probably be stored within a 25 kg sphere. But now the difficulty is how to return the satellite. Extra fuel is required for a controlled braking, and this therefore implies a heavier weight, etc. The starting weight from the earth would therefore have to be at least 1000 tons, probably even more.

At first sight it would seem that the first method would be extremely difficult. The second method, on the other hand, has greater possibilities. A satellite weighing approximately 2 tons, capable of transmitting directed radiation with an effect of 1 kw toward the home planet, would be suitable for direct transmission to the earth of data about the local conditions on Mars or Venus. How would such information be procured?

There are plenty of suggestions about how to do this. If we think about the situation of a robot which has just landed on a hitherto unvisited planet and about what it is supposed to do, we can at once see parallels between this and the experiments that are being carried out in the arctic regions on our own planet to procure automatically information about climate and climate changes. The information organ in this case is a relatively simple device which is parachuted from a plane. On landing, the device unfolds a couple of antenna ribs and immediately starts to relay a series of impulses about temperature, air pressure, humidity, and wind conditions.

In this case the information recorded is, of course, somewhat limited, but a great deal more could easily be included. With regard to the composition of the atmosphere, a portable mass spectrometer can give information immediately about the distribution of different components according to their molecular weight. It is not difficult to telemeter the information to the home base. In principle the transmitter only needs to report the amount of material found per mass number. Something like this:

0	0	0	0	0	0	0	0	0	0	0	0	0	0	0	0	0	11	0	0
1	2	3	4	5	6	7	8	9	10	11	12	13	14	15	16	17	18	19	20
0	0	0	0	0	0	0	7500	0	0	0	2100	0	0	0	0	0	0	0	100
21	22	23	24	25	26	27	28	29	30	31	32	33	34	35	36	37	38	39	40

The results indicate that the molecular weights 18, 28, 32, and 40 are represented by the numbers 11, 7500, 2100, and 100, which—translated into ordinary chemical language—means that

		Per Cent
Water	= 18..........	0.1
Nitrogen (N_2)	= 28..........	78.6
Oxygen (O_2)	= 32..........	20.0
Argon (A)	= 40..........	0.9
Total.....................		99.6

A more complex mass spectrometer would be able to record all the minor components as well, together with carbon dioxide, all the noble gases, and the isotopic components of hydrogen, nitrogen, oxygen, etc. This would give us a mass of data about our atmosphere and its history. The spectrometer can also be used for the analysis of components other than atmospheric ones, both solid and liquid. In other words, the instrument could report to the home base by telemetry that it is on solid ground, that there is such and such a percentage of moisture, and that the main components of the ground surface are distributed according to the following mass numbers: x y z u v a b, and so on. This project is now ready for testing anywhere on our earth. And compact, rocket-borne mass spectrometers have already been tested in the higher strata up to 200 km and have worked satisfactorily.

New and completely automatic gas chromatography equipment

could be adapted, without any great difficulties, for rocket use and for telemetering information. This instrument can be used to characterize certain organic carbon compounds; even the mass spectrometer can be used for such purposes to a certain extent. These two instruments in fact complement each other.

A portable and compact spectrograph for the analysis of the reflected light may also be useful for certain specific purposes.

In the initial stages of exploration these instruments should be able to give us a rough understanding of the milieu in their immediate vicinity. In addition, a miniature television camera could sweep the horizon and also take some close-up photographs of the ground surface. This information about the earth and sky could even be transmitted in color.

The chances, however, of obtaining information by these instruments about the possible existence of any form of life are limited. A robot satellite landing on Mars is quite likely to be set down in a desert area, where there is little to report except the distribution of inorganic substances in the atmosphere and on the ground surface. We ignore the possibility that the television camera might catch, just by chance, something moving. It is probably pitch-dark on Venus anyway, and a television recording taken by flashlight would probably not come up with any dramatic information. On the other hand, the chances of obtaining information about any microflora or microfauna are good. A project already discussed in the United States and originally proposed by the Nobel prize-winner J. H. Lederberg aims at exposing a special type of film in an unknown milieu for a certain period. The film would be coated with a complex mixture of nutritive substances, and after exposure, it would be passed into a chamber where any micro-organisms adhering to it would be able to grow into colonies which would be visible. The growth processes would be followed by a television camera, and the results would be telemetered to the home base. This project, of course, is based on the assumption that all forms of life irrespective of their environment have certain features in common: the ability to live and thrive on certain substances, especially carbohydrates, certain organic acids, and certain amino acids. This subject is treated in greater detail in chapter iv of this book. Technically this project is quite possible, and a preliminary trial here on earth might lead to some stimulating results.

Another experiment worth trying, which complements the one above, is based on the same premise, namely, that the existence of life on a planet implies that some forms will be of microscopic dimensions and transportable, therefore, by wind and weather. I would suggest that an efficient vacuum cleaner be used to suck a considerable amount of atmosphere through a filter, the deposited material then being treated with suitable reagents. The different

TABLE 2

5/27/1971		Wind Speed†	Television Impressions in Daylight‡		Horizon§	Light-Dark Rhythm‖
MDT	Temperature*		Ground	Atmosphere		
1.	+22°	0	3–4	2	1	14–10
2.	−38°	9	9	8–9	1–2–3	3–21
3.	+36	1	3–4–7	2–9	3–4	12–12
4.	+13	2	2–3	8–9	1–2	17–6

* Temperature: °C

† Wind speed: 0 = calm, 10 = gale

‡ Television impressions of color: 0 = black, 1 = violet, 2 = blue, 3 = green, 4 = orange, 5 = yellow, 6 = red, 7 = brown, 8 = gray, 9 = white

components could then be separated by two-dimensional chromatography and visualized with the aid of stains giving certain characteristic colors. If a living form were to be captured, the chances are great that one of its components will represent the group of amino acids, the most general element in the structure of living organisms. The method I suggest presents no insuperable technical difficulties, and its results could be studied by a color television camera when it was free. This somewhat destructive method would not provide information about the way of life of these organisms; but a recorded observation of one or more amino acids in the atmospheric dust from Mars or Venus would suggest the presence of material derived from organisms—either alive or dead—with a probability of 1000 to 1.

Further advances in the capabilities of automatic milieu data transmitters—let us call them MDT for the sake of brevity—would open up many possibilities. They can easily report position data, long-term periodicities, local fluctuations, and a mass of other data, wherever they are placed. It might be amusing to compare the signal patterns of MDT's which have been located for trial purposes in different undisturbed areas on the earth (see Table 2).

It is clear from Table 2 that two of the MDT's have landed in a milieu where the mass number 18 is the dominant factor. The one

Dominating Mass Numbers in the Atmosphere	Mass Numbers, Ground Surface	Organism Test#	
		Growth	Amino Acids
28, 32, 40, (44)	31, 40, 55	1	3
18, 28, 32, 40, 44	18	0	0–3
18, 28, 32, 40, 44	18, 31, 44	10	10
18, 28, 32, 40, 44	18	8	6

§ Horizon: 0 = unbroken, 1 = little contours, 2 = faint contours at a distance, 3 = clear contrasts, 4 = obscured contrasts

‖ Light-dark rhythm: hours of darkness-hours of daylight

Organism test: 0 = none, 10 = abundance

with the low temperature, MDT 2, records night darkness; MDT 4 at +13° C records a continuous blue-green. They are both probably surrounded by water, for the mass number for H_2O is 18. MDT 2, therefore, is located in the midst of snow and ice; MDT 4 is floating about in half-cold water. Both record positive results to the organism test. The night-day rhythm suggests that MDT 2 and MDT 4 are located at opposite ends of the globe and moreover in opposite hemispheres. The revolution time is approximately 24 hours. MDT 1 and 3 are evidently located on firm ground, MDT 3 probably in a moist environment to judge from the mass component 18 which it reports in the ground surface as well as in the atmosphere. MDT 3 records the existence of organisms and different amino acids. The components 28, 32, and 40 are similar over the whole planet, without local fluctuation. N_2, O_2, and A? Nitrogen, oxygen, and argon? 44 is presumably carbon dioxide. The ground surface seems to vary. MDT 2 records a contour at the horizon; is it a mountain? MDT 1 reports an unbroken horizon; MDT 3 evidently cannot see anything in the far distance. MDT 4 can see, but undistinctly, suggesting that there is some kind of haze. Certain rapid local fluctuations appear on the horizon of MDT 4. For MDT 3 local fluctuations occur at intervals.

Each of us can draw further conclusions from these tables about the differing milieus on this planet judging from the reports from (1) the Gobi Desert, (2) Ross Island in Antarctica, (3) offshore from Manáos in Brazil, and (4) the middle of the Sound between Sweden and Denmark.

Such is the technical background of the first attempts to reach the surface of the neighboring planets which have hitherto not been visited by humans. We can see that it is worthwhile reconnoitering at first with MDT robots, even though the data they could procure will probably be strikingly meager compared with what a living observer could report from the same region. A human weighs, say, 80 kg. What would a highly complex automatic data recorder weigh when all the additional equipment required to simulate human perception and comprehension is added? Probably as much or more. But with a human observer plans have to be made for the return journey, with all the extra equipment and fuel that that would involve.

Another problem is how to support one or more observers. Both on the journey there and back and during the period of observation, an observer would be quite beyond reach of immediate supplies from earth. This problem is basically more important than the problem of how to get there and back, which will eventually be solved by the use of a greater initial starting force. This will make it possible to conserve enough energy (or fuel) for a well-controlled return journey. But as far as the support of human beings on other planets is concerned, we cannot rely upon a simple multiplication in the same way; a radically new solution is required.

An adult person under strenuous conditions can survive on 500 g of solid food per day, if the food contains 3000 calo-

FIG. 3.—A highly impressionistic drawing of a data-collecting robot on another planet. In certain cases where continuous telecommunication with the home base is desirable, for example, from Mars to the earth, it may be better to relay the information from the ground surface to a satellite in orbit by ultrashort-wave radiation. The satellite would then pass on the message to receivers on the earth. Equipment to do this is now being tested in various places on earth.

ries; but he also requires at least 500 g of oxygen and 500 g of water. In other words, before nutritional substances can be utilized they require double their weight of oxygen and water. Food for two people for a three-hundred-day trip—a shorter time should not be considered—would involve about 300 kg of food plus 300 kg of oxygen and 300 kg of water—that is, one ton. Furthermore, approximately 500 kg of material to absorb the carbon dioxide would be needed, that is, absolutely essential. The total weight, therefore, would be: 300 + 300 + 300 + 500 = 1400 kg. With a ratio of fuel to useful weight of 200, an additional 280 tons of fuel would be required to launch these essentials. Sending compressed oxygen in closed containers adds to the total weight, which in turn leads to an even higher fuel consumption at the start.

The total weight could, however, be reduced by a number of tricks. First oxygen and water could be transported together as 100 per cent hydrogen peroxide (H_2O_2), which does not need to be stored in heavy containers and which can be broken up easily and effectively according to this equation:

$$\underset{\text{Hydrogen peroxide}}{2H_2O_2} \rightarrow \underset{\text{water}}{2H_2O} + \underset{\text{oxygen}}{O_2}$$

$$680\ \text{kg} \quad 360\ \text{kg} \quad 320\ \text{kg}$$

Moreover, certain peroxides, for example, lithium peroxide, Li_2O_2, are good absorbents for carbon dioxide, the absorption taking place at the same time the oxygen is released from the peroxide:

$$Li_2O_2 + CO_2 \rightarrow Li_2CO_3 + \tfrac{1}{2}O_2$$

$$460\ \text{kg} \quad 440\ \text{kg} \quad 740\ \text{kg} \quad 160\ \text{kg}$$

As we can see from this series of formulas, the absorption of one mole of carbon dioxide gives rise to one half mole of oxygen; during a three-hundred-day period, the carbon dioxide produced by two individuals would correspond to approximately 160 kg of oxygen, or half the amount needed. However, the two series of formulas can be combined to some kind of compromise:

$$Li_2O_2 + H_2O_2 + CO_2 \rightarrow Li_2CO_3 + H_2O + O_2$$

$$460\ \text{kg} \quad 340\ \text{kg} \quad 440\ \text{kg} \quad 740\ \text{kg} \quad 180\ \text{kg} \quad 320\ \text{kg}$$

Here the production of carbon dioxide and the consumption of oxygen by the travelers equal the carbon dioxide absorption and oxygen production from the two peroxides. The total weight of hydrogen peroxide and lithium peroxide is in this case 800 kg, which, added to the dry food, gives a total weight of 1110 kg. This arrangement eliminates about sixty tons of fuel in the starting weight and allows us moreover to obtain 180 kg of water, exactly half the amount required for the whole period without rationing. The other half would be obtained by utilizing the water we carry inside ourselves by reusing excreted water. As a matter of fact successful experiments of this type, in which experimentalists have lived on their own water, have been carried out. Water from both respiration and urination has been collected, purified by distillation and special chemical treatment, and redrunk. In the United States, especially, these experiments have been perfected, and this is of considerable importance for the nutritional problem.

All these calculations represent but a minimal fraction of the multitude of considerations to be taken into account when launching a manned spaceship on a long-distance mission. At the time of a manned round trip to Mars—let us guess in the early 1980's—compact nuclear power units within the ship will be operating not only for the purpose of course correction but also for the general task of carbon dioxide disposal and perhaps also for food production on a limited scale. The latter area imposes problems of tremendous difficulties but nevertheless has already been tackled in many research laboratories.

A promising solution which at present is being eagerly investigated is a closed system which includes both people and green algae. The former produce carbon dioxide and water and consume oxygen; the latter consume carbon dioxide and water and produce oxygen. These two systems balance each other, provided that a sufficient amount of light energy is pumped into the algal system to make it function as a carbon dioxide assimilator:

$$CO_2 + H_2O \xrightarrow{\text{light energy}} (CH_2O) + O_2 \quad \text{growing algae}$$

$$(CH_2O) + O_2 \longrightarrow H_2O + CO_2 \quad \text{consumers of food and oxygen}$$

These elements combine into a miniature of life in which the animals consume the oxygen produced by vegetation under the stimulus of sunlight. The cheapest solution would be to build windows in the spaceship large enough to let in sufficient sun energy for the algae. Yet some artificial light would most probably have to be used, which would of course use up much-needed power. This system is now being tested for use in nuclear-powered submarines as well as for flights to the moon. For long-distance journeys the system may have its advantages. For an actual landing on an uninhabited planet, it is an absolute necessity.

This technical discussion ignores the human aspects of the problem. But these experiments are part of what will go on in the next few years before man makes his first attempts to visit his neighboring planets; they are preludes to the construction of the observation bases which will form the first stage in a colonization movement. Robot probes are a highly important element in these projects, and the experience gained from these probes will determine to a large extent the future program. Innumerable technical problems need to be solved before manned space flights to other planets can take place, especially the problems of satisfactory maintenance and the establishment of supply routes.

The moon will undoubtedly be the first object of expansion. This is because of its proximity, and it will become a testing ground for exploration further afield. But these longer journeys, especially to Mars, may take place sooner than we now anticipate. Within the neighboring planet system Mars is the planet with the most favorable environment. The temperature is fairly tolerable; water, although its content in the atmosphere is much less than on earth, is most probably present to such an extent that a landing expedition could utilize the local water supplies with the help of some technical tricks. The external atmospheric pressure is considerably higher than on the moon, possibly around 10 to 20 mm. The gravitation is favorable for the construction of bases as well as for making return journeys. And there is one problem to be solved: whether endogenous life exists.

Venus is a less attractive proposition in many ways. The latest data we have about the ground temperatures indicate that the climate is a great deal hotter that it is in the tropics. The extremely dense cloud cover presents great technical obstacles to the construction of bases and observation points. On the other hand; what surprises might be there for interested chemists?

Farther away from the earth the goals become increasingly less attractive, and journeys to these cold regions with their temperatures of —120° C and lower will probably take place much later and then hardly with a view to colonization but perhaps to the establishment of an outpost toward emptiness.

In order to clarify our objectives, let us consider what work and everyday life will be like on the moon, Venus, and Mars when the first manned expeditions land on these planets. What we can determine about the environments and the conditions of life should be taken as extrapolations of our present experience, as illustrations of man's ability to adapt and recreate. The following section about the exploration of our planetary system is based as far as possible on fairly reliable data; the more subjective parentheses represent man's reflections on the ability of nature to vary milieu and conditions of life, and also perhaps man's feelings at some moments of great revelation.

CHAPTER III

FROM THE PERSPECTIVE OF THE MOON

The first landing maneuver of a manned expedition to the moon will present some interesting problems. Even if a great deal of information about the ground structure has been gained through preliminary robot landings and telemetered in great detail to the home base, the moment of contact with the ground will be one of considerable risks. The ground surface is probably highly variable in the different regions of the moon, for there are extreme differences in altitude in the mountain regions, and even a level surface may prove a treacherous foundation for a soft landing. A spaceship cannot be maneuvered easily and does not have the ability to choose in any detail a suitable landing spot. Moreover, the gas jets of the braking mechanism will stir up an immense cloud of dust even at a height of some one hundred meters or so above ground, and this will impair visibility. Still, this is all in the game.

It would be a great advantage if the ground within a radius of one or two hundred meters could be cleared. When the dust had settled—this does not take long in a vacuum—the most suitable landing spot could then be selected, taking into account hills and crevices. Just before landing one or two sounding projectiles with enough power to sweep the area free from dust and gravel should be launched. We should view a level surface on the moon as something which might give us all sorts of trouble.

The landing legs of the spaceship yield at the first moment of contact, the vibrations of the motor fade away, strain gives way to relaxation. Relief. So far. . . .

After a pause to catch one's breath and to exchange compliments with the earth base, the first reports must be made. They will not contain anything of fundamental importance that we do not already know from the robots sent up earlier. What can be seen through the plastic observation windows is already common knowledge, for the findings of television cameras have been telemetered to the earth for many years. The topography of the hidden part of the moon is now well known; we expect the violent contrasts between the pitch-black shadows and the burning luminosity of the reflected sun, the sharp contours of the mountains on the far horizon, the dust banks of the ground surface, and the impression of sterility and quiet intensity. We expect and find a cosmic desert.

The detailed working program allows only a few seconds for the contemplation of nature. Someone probably glances quickly into the intensely black space in the hopes of catching a glimpse of the opaline ball that is earth, where so much preparatory work has been done during the last decades. Maybe there is a nostalgia for what seems an idyllic time, the 1960's, when the data about the moon were acquired by observations from a distance, a time before robot inspections, a time when man's knowledge of lunar chemistry and physics was essentially rudimentary and based on probability calculations. A time when data could be summarized quite simply in tables as starting points for extrapolations which were often quite naïve. The past. The planning stages.

A summary of data about the moon compared with the corresponding values for the earth is given in Table 3.

This table gives us a basis by which we can judge the environmental factors on the moon. Topographic data have already been published in maps which have been compiled over the years in many different parts of the world.

By measuring shadow lengths and their variations during a moon day it has been possible to estimate approximately the altitude of the highest mountain peaks. The Apennines—a mountain chain close to 1000 km long—have peaks higher than 5000 m.

Mount Huygens, rising about 6500 m above its surrounding landscape, is the highest peak in this range. There are mountain ranges near the South Pole, in which a couple of peaks, Leibnitz and Dörfel, reach the impressive height of more than 10,000 m, considerably higher than Mount Everest with its 8892 m.

In this mountainous world where no wind has blown for the past five billion years and where weather means only temperature changes, erosion has taken forms other than those we are accustomed to on earth. The temperature changes, with a range of over several hundred degrees, must have caused a gradual formation of crevices in the bedrock—in the bedrock itself as well as in the form of crumbling rocks and loose boulders. Because of the low gravitational force, the eroded material does not always fall away. Mountain climbing would have to be undertaken with caution because of the loose stones and boulders which for thousands of

TABLE 3

Data	Absolute Values		Relative Values (Earth = 1)	
	Moon	Earth	Moon	Earth
Mass	0.0735×10^{27} g	5.975×10^{27} g	0.0123	1.00
Radius (mean)	1737 km	6371 km	0.273	1.00
Density	3.34 g/cm^3	5.52 g/cm^3	0.605	1.00
Orbit radius	149.1×10^6 km =	149.1×10^6 km	1.00	1.00
Period of revolution about the sun	365.256 days =	365.256 days	1.00	1.00
Period of revolution about the earth	27 days, 7 hr, 43 min			
Gravity	161 cm/sec^2	981 cm/sec^2	0.165	1.00
Escape velocity	2.4 km/sec	11.3 km/sec	0.212	1.00
Albedo	0.07	0.29	0.24	1.00
Mean temperature	+4° C =	+4° C	−150° C to +150° C	−80° C to +80° C
Distance between the bodies	384,400 km		(local variations)	

years have been balancing precariously on the slopes and shoulders of the mountains. The slightest interference could cause a tremendous avalanche.

The angle of fall for loose gravel and sand is difficult to calculate, but probably it is much steeper than that on earth. On the moon, however, a fall would be quite different from one on earth, because boulders, stones, gravel, and dust would fall at the same rate in the vacuum, a point which ought to be taken into account on excursions. Caution is always advisable even though a man with his weight reduced to about 10 kg, and protective clothes weighing 15 kg, can climb Leibnitz with no greater effort than climbing in the Catskills. It is true that a boulder measuring half a meter across weighs only one-seventh of its corresponding weight on earth, but after all, 50 kg is 50 kg, especially if it has gained speed. The gravitational force per se is one-seventh of that on earth, but the inertia is the same. The acceleration or deceleration of an object requires the same force as on earth, so the dangers of falling boulders and stones should not be underestimated.

According to older views the ground and the mountain ranges are formed of weathered volcanic lava and degassed molten basalt, the results of an earlier, large-scale volcanic activity. This theory, of course, is derived from the similarities in shape between the form of the mountains on the moon and the volcanic craters on earth. But there are differences of opinion on this subject. The research group to which the American Nobel prize-winner H. Urey belongs considers that the ring formations have been caused by the impact of meteorites over a period of billions of years and that the surface of the moon has never in fact melted.

The supporters of the volcanic theory point out the difficulties in reconciling the great size of some of these crater mountains—up to 200 km in diameter—with the damage which could be caused by the impact even of very large bodies from outer space. At the moment, as far as we can judge, the volcanists seem to be getting the better of the argument. The English scientist V. A. Firsoff has made a very close study of this problem, and he concludes that the shape of the ring mountains clearly reveals their volcanic origin and that the folding of the mountain chains and the other tectonic phases on the moon follow by and large the processes we have already observed on earth.

One interesting theory that has come out of Firsoff's work on the moon's surface suggests an extremely thin atmosphere around the moon. This contrasts with earlier theories that the moon's surface is in a complete space vacuum. It is clear from the low attraction power of the moon that there can be no high density gas. This theory is also supported by the low value of 2.4 km per sec for its surface escape velocity and its thermic past. All these explain the inability of the moon to retain to any large extent both existing and newly produced gases.

Firsoff seizes upon the fact that volcanic ash and minerals have on the whole a high adsorption capacity for different gases. He hypothesizes that a small but significant fraction of the original ancient gases may still be trapped in the surface material. The heating of the surface layer by the sun during the moon day implies a liberation of a certain amount of gas, and the lowering of the temperature at night, a renewed adsorption. If this is so a constant amount of gas is diffused from the warm to the cold parts of the surface of the planet creating a gentle breeze. The amounts of gas are not, of course, large; a pressure of 0.07 mm near ground level or perhaps a little higher has been put forward as a plausible figure, that is, one ten-thousandth of our own air pressure at sea level. Anyway, Firsoff's theory has many points in its favor, and it does now seem certain that the moon surface produces gases to a limited extent.

First, argon-40, a product of the radioactive decay of potassium-40 must be produced on the moon as well as on earth. A simple calculation reveals that the argon content of our atmosphere is increased by about 10,000 tons per year. If we estimate that the moon's surface is one-sixteenth as large as the earth's and assume that the same amount of potassium is present in the surface rocks —a highly reasonable assumption—we end up with a yearly production of argon on the moon of approximately 700 tons. The atomic weight of argon is 40, which means that it belongs to the medium weight gases, and so it would have a good chance of being retained at least to a limited extent. It can also be adsorbed by porous minerals, a further reason for its retention. On the assumption that the moon has retained all the argon produced in the bedrock since its earliest history, the American astronomer H. Shapley has estimated that this would result in an outer pressure corresponding to 0.7 mm by our measurements, that is, ten times greater than the figure given above for the ground pressure.

At this point the question of water and carbon dioxide comes into the picture. These substances are adsorbed in certain minerals at respective rates of 30 and 10 times higher than argon. Some silicic acid structures have a very high adsorption capacity for water even at temperatures far above those of the moon during the day (approximately 150° C). Low outer pressure decreases adsorption capacity, but calculations based on experiments show that the amount of water bound in 1 kg of silicic acid structure at 0° C and 0.1 mm pressure is 50 g! This is, of course, a special case and the natural binding capacity for water in the lunar ground material is probably only one-tenth of the above capacity, perhaps only one-hundredth. But after heating it to a temperature of 500° C or so, 1 kg (earth weight) of ground dust from the moon ought to yield—even by the most pessimistic calculations—0.5 g of water. That water comes into this discussion at all is the result of the simple fact that all geologic processes, from the formation of granite to more dramatic events such as volcanism, liberate chemically bound water to a large extent. Our own volcanic gases consist of more than 90 per cent water vapor.

The question of lunar water has another aspect. During the early history of the moon before its surface cooled to its present mean temperature, that is, at least four billion years ago, there must have been a prolonged period of volcanic activity and as a result water vapor must have been released. Most of this was probably soon dispersed, but there is a chance that some water still exists even in fossil form from this early period. There are several places on the moon's surface which, because of their topographic location, are never reached by the sun and have a constant temperature of —150° C. At this temperature water has a vapor pressure of less than 0.0005 mm. These places, therefore, are favorable centers for the collection of volatile substances. The possibility of actually finding fossil water and even carbon dioxide on the moon should not be underestimated. This would be of enormous importance for the maintenance of an observation base.

Has the volcanic activity on the moon ceased altogether? Are gases still liberated on a small but perceptible scale? But what do we mean by volcanic activity? Large-scale eruptions complete with lava-streams and so on undoubtedly belong to the past. There could well be small sporadic emissions of gas from hidden reservoirs, especially in areas covered by thick dust. It has been calculated that the dust covering the moon's surface, apart from being a good adsorbent for gases, should also be an extremely poor conductor of heat and should act as a kind of insulator. Some research scientists are of the opinion that the temperature a few yards under the dust layer, at a depth of perhaps a foot or less, should be relatively constant and unchanged by the fluctuating temperatures at the surface level. In their view it would be something like the permafrost layers in the earth's polar regions with their fossil ice. Frozen volatile substances, especially water, can exist under these conditions to a limited extent. In rare cases, however, the heat of the day period would cause an increase in the temperature of the underlying strata with some vaporization as a result. On rare occasions gas might break through the top of the dust layer and scatter gravel and dust in a momentary eruption. It is a sign of something that is not quite asleep. A cloud of dust rapidly settling is its only visible sign. A few seconds of activity, then nothing.

A large number of temporary mists or local veils of smoke have been observed in isolated regions of the moon. They are extremely rare phenomena, but only recently has much attention been given to them. Interest in them links up with research on the problems of the lunar atmosphere.

A possible summary of the atmospheric components on the moon, with certain reservations about the accuracy of the data, is given in Table 4. This table needs a few comments. Astronomers of today agree that the moon *has* an atmosphere but that it cannot be very much denser than one ten-thousandth of our atmosphere. The value 0.01 mm is in this order of magnitude; 0.1 mm may be too high, but it cannot be completely disregarded. The argon content problem has not yet been fully studied, but I find reasons to believe that argon is the dominating component. However thin the moon's atmosphere may seem to be, it nevertheless has a property which is not always taken into consideration. It should offer some protection to explorers from bombardment by micrometeorites and some braking power for spaceships, which is important from the fuel economy point of view. The density of the atmosphere at a height of 50 km over the moon's surface corresponds to the density of the earth's atmosphere at a height of 100 km because of the earth's higher gravitation, which causes the density of the earth's

TABLE 4

Components	Pressure (In Mm)*	Total Amount over the Whole Surface (In Tons)†
Argon	0.07 -0.7	10^9–10^{11}
Water vapor	0.02 -0.1	10^8–10^9
Carbon dioxide	0.015–0.02	10^7–10^8
Sulfur dioxide	0.005–0.02	10^7–10^8
Nitrogen	0.001–0.005	10^6–10^7

* Total pressure = 0.1–1.0 mm. The lower figure is the safer one.
† Total approximation: . . . 0.00001–0.0001 atm . . . .

atmosphere to decrease more rapidly than that of the moon. The braking power provided by the lower strata of the moon's atmosphere should not be underestimated and the advantages of circling the moon many times in order to save precious fuel for the final vertical landing maneuver cannot be overestimated.

Finally, it is possible that the moon's atmosphere may have an ionized layer similar to that of the earth's atmosphere which can reflect medium- and long-wave radio waves. This would have important consequences for future communications between different bases on the moon and for telecommunications between explorers way out in the desert. If there is no ionized layer the only possible way of communicating on the moon would be by short wave and ultrashort wave which would lead to a great many practical difficulties. The moon has a rather hilly topography and short-wave communications in the 50 mc range have only an optical scope and are therefore completely screened by obstacles such as mountains. Two bases on the moon would find that the easiest way to exchange information would be via the earth. A more complex solution would be to use satellite balloons kept in a definite orbit around the moon, such as the American Echo, in order to establish communication with someone—perhaps only on the other side of a valley—by reflexes from the metal-covered surface of the balloon. If the first robot probings could demonstrate the existence of an ionized atmospheric stratum around the moon, this would save a great deal of trouble.

Our experience of building in a vacuum is somewhat limited. Those who attempt it for the first time will be faced with a situation which is impossible to simulate on the earth by experiments. Those pioneers whose orders will be to remain at the landing point for a certain period will be taking part in a premiere. Their first consideration will be to make use of the terrain and locally available materials for construction in order to cut down the number of things that will have to be brought from earth.

The spaceship could hardly be converted into an observation base, for its interior would have been most carefully planned to allow only a minimum freedom of movement. This is essential for the journey, but it means that any longer stay aboard after a landing is not very convenient. But the ship will be the natural starting point for the local construction program.

One of many suggestions is that a large, thin plastic cylinder should be blown up like a balloon out of the ship so that it forms an annex to the spaceship, with free access from one to the other. Then comes the hard work, for shoveling dust with the added weight of protective clothing is a laborious business, but one for which the explorers would have already been trained at home. A hollow into which the plastic cylinder can fit snugly with many layers of dust over it and even more dust as an insulation layer, and even more, and *voilà*—an igloo. A half to one meter of dust and gravel over the cylinder would probably be enough to protect it against the heat during the day and the severe cold at night. An igloo, strengthened possibly by one or more inflated inner envelopes, may become the first primitive dwelling on the moon—the starting point for more permanent buildings.

A common feature of any permanent settlements on the moon, at least during the initial stages of colonization, would be their subterranean location. The neat residences of science fiction with their cupola roofs would depress any very realistic architect. It is necessary to have an inner pressure of one atmosphere compared with the vacuum of the surroundings, a parallel to what we would need to achieve on earth in order to keep an excess pressure of one atmosphere in a summer residence. The shape will have to be that of a cylinder or a sphere, like the storage containers for thousands of cubic meters of gas under a pressure of a couple of atmospheres. In order to start such a building at all, construction material will be needed. In the absence of material one digs a cylindrical hole, either vertically on level ground or, better, horizontally into a mountain side. The hole is then filled by an inflated balloon which is strengthened by a coat of plastic spray. These balloon-filled holes are then joined by similarly coated passages. A network of passages is built to join these holes. Larger and larger caves are built for equipment. A minimum of material has been brought from the earth; there has been a lot of digging, drilling, and blasting. The first moon bases start to take shape.

The maintenance of an observation base on one of the neighboring planets—the moon, to begin with—will cause some unexpected problems. The first construction period will be the most critical time for the pioneers on the moon. The spaceship that has brought them and the essential equipment cannot be expected to carry as well extra construction material, oxygen, or food, etc. The best suggestion that has been made is that this extra equipment should be stored in orbit around the moon: as circling packages which can be controlled. The pioneer group could then release the braking devices of one package at a time; this, of course, would have to be done at exactly the right moment under the right conditions; it would be a task for an electronic data machine with a programing device which had been brought from earth. The successful realization of this operation is of vital importance to the project. A delivery of explosives, oxygen, or food that lands, because of a miscalculation, some 1200 km away will be of little use to someone who needs to build a place in which to live, to breathe, and to eat right away. Experiments in the United States have shown that the retrieving of space-borne containers is possible. The difficulty about using this system on the moon is that a circling container could not be reached by an order signal beyond the horizon, that is, beyond the scope of vision. On the moon the horizon is much more limited than on earth because of the more pronounced curvature of the moon's surface. There will probably have to be a very exact co-ordination between the earth systems and the transmitters on the moon before all the supplies can be landed in the immediate vicinity of the base. To everybody's relief.

In the long run all imports imply an activation of the local economy, for food, power, and building materials can be developed. As far as food is concerned, there is only one possibility, that is, the development of local photosynthesis, as we have suggested earlier. The quickest way to organize this project would be, of course, to set up greenhouses in caves in the mountains where they would be relatively insulated from temperature changes. It would be impractical to build them in places which are directly exposed to the sunlight. One solution that has been suggested is to cover the angle between the mountain sides and the ground with a stable transparent cover; the sun's rays could then be directed onto these areas by mirrors, and a simple turning device could make it possible to direct light energy into these protected greenhouses at all times regardless of the position of the sun.

The shape of the greenhouses, the culture media, and the types of plants to be grown will give botanists and agriculturists many problems, but they should not be too difficult to solve. First, the sand and dust of the moon surface will probably prove an excellent soil for cultivation in general, especially since no water during the last billions of years has soaked through it. The choice of suitable plants will be more difficult. The plants will have to be extremely reliable; the entire well-being of the observation base in terms of food and oxygen supply and air purification will depend on the operation of the greenhouses; a crop failure would be a disaster. The plants would have to be effective from all points of view; they would have to be fast-growing, edible, and capable of survival during the recurring cycle of a fourteen-day interval of darkness. The latter represents a real problem and requires a careful selection of flora. A possible solution would be to lower the temperature in the greenhouses during the intervals of darkness in order to inhibit the respiration of the plants. This would mean that fourteen days of summer activity, growth, and oxygen production would be followed by fourteen days of semiwinter with a partly frozen chemical cell activity. It would be an interesting task from the point of view of plant physiology to select plants capable of living and thriving under such a rapidly changing summer-winter rhythm. One solution that seems obvious, the use of artificial light during the dark period, is impractical because the electrical power supply will also be dependent to a large extent on sun energy. Highly efficient accumulator systems would be required to provide lighting for both the greenhouses and the dwellings during these dark intervals.

A compact nuclear power plant will be among the first items of equipment, and it will probably be sent at an early stage by orbit. Under these circumstances such a power plant would not need any heavy or bulky protective cover to screen radiation and so would not need to be particularly heavy. To eliminate any risks, it could be sent in an inactive form and "aroused" at the destination. The pioneer group could lower this device into a crevice, connecting it to the base by cable so that it could be started and controlled from a distance. Starting and supervising it would soon become a routine matter. The nuclear power plant, however, would probably be the reserve source of energy for the base, for sun energy would prove the most economical in the long run. Small sun power plants have already been tested in laboratories, and these may prove to be of great value on the moon, where the weather during the day period is absolutely reliable.

The first moon base gradually evolves into a primitive form of co-operation between the vegetation and the animals, that is, the organizers. This reflects the way in which life is organized, though on a larger scale on our earth—an application of principles which we shall discuss in the next chapter. We can see the fundamental idea at work in a simple technical form in Figure 4.

FIG. 4.—A design for an observation post on the moon may look something like this. Incoming sunlight is reflected by revolving mirrors into the oxygen-producing units, which are sealed off from the vacuum of the outside world. By building the base in the mountain side, it should be possible to obtain a fairly even temperature, which could be maintained by heat obtained from sun generators or from an atomic power plant. Theoretically, it is possible to achieve a balance in the system, but this could only be brought about after the expenditure of a good deal of time and effort. Once an equilibrium was achieved, the base would be self-supporting as far as food and air are concerned. The inevitable flaws in the system could be compensated for by a reserve stock and by further supplies from the home planet.

In principle the system is self-supporting as far as oxygen, water, and food substances are concerned. But this is a somewhat idealized system which cannot, of course, function autonomously in the long run. Every now and then it will require stimulus in the form of new seeds, fresh supplies of water, carbon dioxide, and civilizing amenities for the personnel. As far as water and carbon dioxide are concerned, it has already been pointed out that the surface material of the moon might possibly yield a certain amount of water on thermic heating to several hundred degrees and presumably also carbon dioxide. The development of solar energy power plants might activate such processes for the concentration of lunar water and carbon dioxide. It ought to be possible to supply 1 to 10 kg of water per ton of treated material, which would be about the amount needed to compensate for the inevitable leakage from the organization as a whole.

It is obvious that the developments which we have outlined here could take completely different forms. Plant material, which in this context suggests, unless it is further defined, a process of continuous consumption by a permanently hungry personnel, can be varied so that the diet is not too limited, either from a physiological or, more importantly, from a psychological point of view. A year-long diet of sugar beets, for example, would become very trying after a while, even with the use of great culinary expertise. A little biotechnical ingenuity, for example, by the addition of a few rabbits and chickens—which could be done quite easily—would break the monotony. A modest brewery would be feasible and is to be recommended as a purely psychological stimulus which would increase the activity and the inventiveness of the personnel in the organization. All these developments are, however, substrate or oxygen-requiring and would demand a corresponding increase in the plant material, a technical illustration of the complex networks of co-operation patterns between the ever increasing forms of life on our own earth.

This whole discussion should be thought of as an introduction to the biochemical lines of inquiry that follow. A question worth considering in connection with the future colonization of the moon is whether any kind of life could be organized outside the closed unit, that is, on the moon's surface itself, exposed to the brutal changes between excessive cold and excessive heat. This question is often brought up in literature on the moon and it would be unwise to dismiss it as unrealistic without any further investigation. The problem depends upon the integration of the following factors:

1. Access to water and carbon dioxide
2. Access to sunlight
3. Ability of the organisms to withstand the local heat of the sun
4. Ability of the apparatus for absorbing water and carbon dioxide to function in excessive cold
5. Ability of the organisms to economize on water in a milieu of low external pressure

An integration of these factors to some kind of unified organism would produce a kind of eclectically shaped lichen-cactus of exotic appearance, to say the least. It should have an actively metabolizing root system capable of absorbing either fossil-bound water or carbon dioxide from sheltered, tempered stores of these components and the ability to use sunlight—preferably reflected—for the photocatalyzed formation of organic material. Furthermore, it must have the capacity to counteract the high degree of evaporation which would exist in any system exposed to an atmosphere which is almost similar to that of a regular laboratory vacuum. To breed such a plant would be a worthwhile project for ambitious plant improvers. It should be technically possible—we must disregard the wilder products of the imagination—but it would require a higher level of biological knowledge than we have at present. The critical factor in such a project is, of course, the extent to which fossil water actually exists on the moon, and the extent to which it can be mobilized by a living organism. As to the theory that something of this kind already exists in the sheltered ecologic niches on the moon's surface, we had better wait and see. This subject comes up again in the discussion about life on Mars, which is touched upon in chapter v.

This outline of the first phase of expansion outside the earth is only a prelude to the realization of more ambitious tasks. The establishment of a lunar base as an autonomous unit would lead to a further dissemination of analogous units to other locations on the moon, necessitating the establishment of communication lines and a further utilization of the planet's local resources of solar energy, minerals for construction material, and atomic energy. Industrial development would be the next phase, then the development of the network of material, power, and information exchange that we call civilization. The particular forms of this program and the political and economic problems that will arise in such a development are, however, outside the scope of this work.

Apart from provoking these technical developments the moon will serve as a base for astronomic research and for an attempt to analyze the structure of the universe from a new perspective. The low gravitation on the moon favors the much more efficient construction of highly developed instruments than is possible for us at present; for example, a giant reflector telescope could be built with a larger and more stable bed than is possible on earth. The information gained from these instruments, both about the cosmos and our local planetary system, will be of vital importance in any further expansion, particularly in the exploration of Mars and Venus and the possibilities of encountering endogenous life there and in obtaining information about the outmost limits of the universe.

Extract from a letter from Base 2, August 14, 1979:
. . . and what two years ago seemed to be work purely for an ideal —and how we worked—is now becoming something else, at least for me. The routine is beginning to get me down and daily contact with the other members of the team is tedious after a while. They are fine people; there is nothing wrong with any of them; but I find that being confined to one place all the time is unbearable, and now when I am free I often go off and work on my own. I feel I need time just to think things over and get everything in perspective, get a grip on all this in some greater context—you know. Everything has happened so quickly I haven't been able to keep up with it all. I have worked on it as a specialist, but I

have no real contact with the project as a whole. We are physically in good shape; the food is all right, but it is monotonous, even though we take turns trying to think up something new. Do you have any suggestions—something quite revolutionary—about how we can make our algae, beets, and yeast extracts into something new? Martinez, who is the expert, says that he can't think up anything more. The boss is against all experiments with animals as an intermediate stage, so that is the end of that. A guy who raised a couple of lab guinea pigs secretly in the greenhouse and had a little private meat orgy got one month's hard labor in the purification plant.

I've been working away on my own on astrophysics and biochemistry to try to broaden my general knowledge. Nuclear physics and technology keep me busy enough during the day shifts, so I need something new to keep me entertained. When I get home, after this contract is over, perhaps I can apply for a job in another project. There's so much on the move right now. Where? I have not the slightest idea, I shall just have to wait and see.

There is earth light outside, and I had the same old feeling of belonging to the great whole when I went out a short time ago to inspect the reflector mechanisms (they are working fine, by the way). The boss was also out for some reason or other; we didn't talk much; we just stood there for a while in the darkness. Funny guy, the boss, I wonder what he's thinking about, walking about when he should really be asleep. Well, maybe we have the same problems. But as I've told you before, it's just the way you take it. I wonder what this place will be like in a couple of years. It's an expensive business and signals about rigid economy arrive every now and then from the higher-ups. Earth people have no idea about how things are here on the base. They just sit behind their desks and think, just as they did when they first described how everything was going to work out long before we even got here, and what things were here and what were not and vice versa. Anyway, it's great to be going on an expedition to the caves tomorrow. Ivan is very hopeful about what we are going to find there. He sends his best, and I wonder. . . .

CHAPTER IV

PREPARATIONS FOR AN ENCOUNTER

The most exciting time in the future exploration of our planetary system will be, from many points of view, the first ground contacts with Venus and Mars, with the possibility of encountering some form of endogenous life. During the past hundred years this question has been looked at from many different angles and imagination has often made the most of the situation by conjuring up dramatic presentations of the subject. The remoter the time when an actual investigation could be made of our neighboring planets has seemed, the wilder have been the speculations, so much so that they seem like part of a classic adventure story in particularly exotic surroundings. There have been fictional stories about discoveries of ancient and highly developed civilizations or remarkably primitive ones; but all these stories present pictures of conquest—even the sword-in-hand type—looting, and possibly some very romantic scenes. For years we have been waiting for the chance to land as *conquistadores* on a foreign continent, to find treasure, and to rescue attractive princesses or enslaved beauties—if the worst comes to the worst, even fellow explorers—from every conceivable kind of calamity. A common feature of these stories, however, has been the cool or even openly hostile attitude of the population toward our very justifiable plans to trade, to take over power, or to obtain information.

As we have acquired more data about the most prominent surface features of our neighboring planets, we have had to modify our expectations of finding an advanced and conscious form similar to species Homo sapiens in some distant part of our planetary neighborhood. It appears less and less likely that an encounter between modern man and any planet organism could lead even to an exchange of ideas, not to mention any more penetrating discussions. But, the question is whether we can count at all on the possibility of encountering life on the organism level in the form that it exists on our own planet.

In many ways this is largely a question of definition. What we consider to be life on our earth is the whole complicated expression of a combination of chemical reactions between two main groups of units. The aims of both these groups are opposed as far as the utilization and replacement of the chemical components on the earth's surface are concerned. There is the plant kingdom with its specialized ability to build up, through light energy, complicated organic material from the water, carbon dioxide, and mineral salts in the environment. This applies to all the variations of plant cells in various forms of organization. The oxygen in the atmosphere and the sea is utilized by the second main group of of living beings: the animals. The animals consume both plants and other animals and regenerate carbon dioxide and water. As a result there is a continuous movement of both simple and complex material between the groups, the reservoirs (representing the plant kingdom), the animals, and inorganic material in equilibrium with the former.

Within this framework, where oxygen plays the dominant role both as something produced by green plant cells and as a destructive agent for animals on the whole, there are certain areas of life in which the energy necessary for the construction of complicated material from simple sources of material is derived not from extraterrestrial sources—sunlight—but from purely terrestrial energy. Let us take an example. Certain sulfur bacteria, isolated completely from the oxygen-activated bustle of the world, live by utilizing hydrogen sulfide as their energy source. This is a parallel to what we see in green plants in their utilization of water under solar energy activation. Another energy source is metal ions at a low oxidation level. This would be similar to the utilization by our primitive iron bacteria of ferrous ions at the +2 level for the active incorporation of inert carbon dioxide into complex carbon compounds (see Fig. 5).

There is a striking similarity between the autonomous systems and the dominating globally feasible systems. The cells and their structure, the components of the cell material, their construction in principle, and the organization of the intracellular enzyme systems are the same. The difference between the systems, apart from their exterior shape, lies in the energy-supplying processes with regard to the utilization of the available energy sources.

Sections *a*, *b*, and *c* of Figure 5 symbolize the synthesis mechanisms in general in which the supply of energy produces an incorporation of simple carbon material into something complicated which is then embodied as a unit in a growing complexity. Carbon dioxide here represents the prototype of simplicity at its lowest level that only needs outside energy to be incorporated into the cell organization. It is a special case, the culmination of a development that probably took place at first in local milieus with considerably more complicated substances as starting materials for the growth and development of primitive cell systems.

There is, however, process *d* by which animal cells take in complex material—food substances and former organism material—which is utilized for the construction of a complex cell organization, the energy being obtained partly from the material itself through enzymatic breakdown, often down to the level of carbon dioxide. with the co-operation of oxygen. Let us make a simple analogy: a business with access to an oil well can use some of the oil in diesel engines and other combustion processes in order to convert the chemical energy of the oil into other forms of energy such as heat and electricity, which in turn can be utilized to manufacture complicated products such as chemical and plastic products from the rest of the oil. Some of these prod-

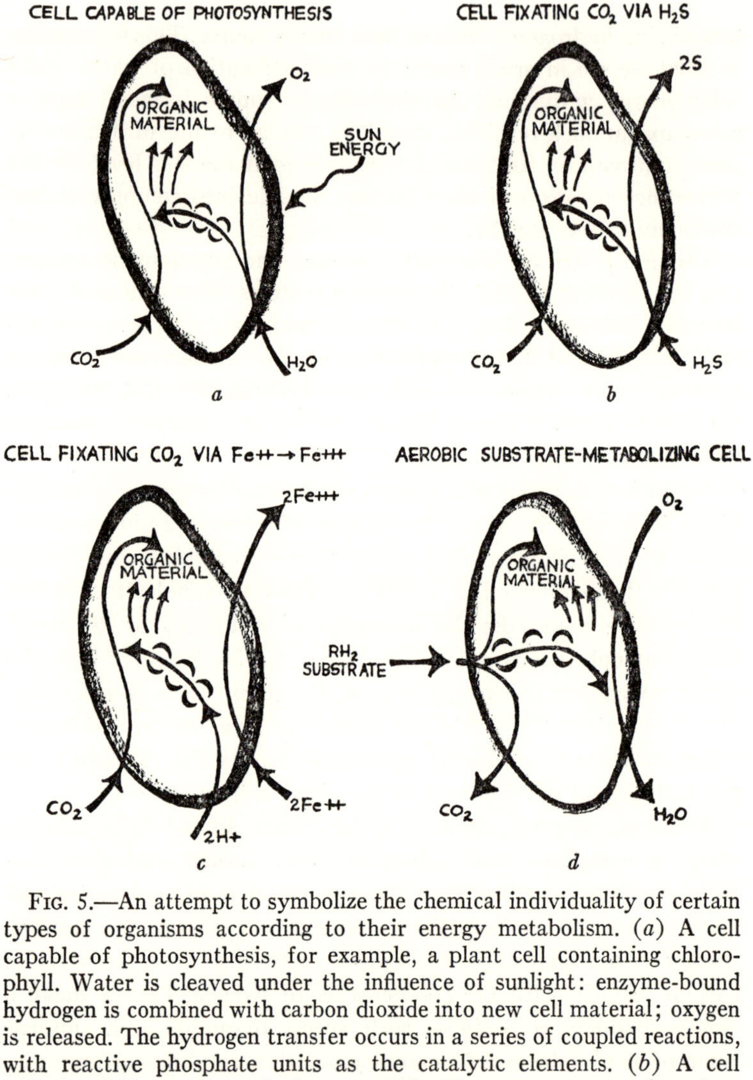

FIG. 5.—An attempt to symbolize the chemical individuality of certain types of organisms according to their energy metabolism. (*a*) A cell capable of photosynthesis, for example, a plant cell containing chlorophyll. Water is cleaved under the influence of sunlight: enzyme-bound hydrogen is combined with carbon dioxide into new cell material; oxygen is released. The hydrogen transfer occurs in a series of coupled reactions, with reactive phosphate units as the catalytic elements. (*b*) A cell with the ability to cleave hydrogen sulfide. The same process takes place as in *a*, except that in this case sulfur instead of the oxygen of the previous system is split off. Cells of this type are represented by certain specialized types of bacteria, such as sulfur bacteria. (*c*) A cell capable of utilizing iron salts (Fe^{2+}) and hydrogen ions for its synthetic re-

ucts can then be used for further development and organization of the business either directly or indirectly.

If we want to carry the idea further and make a similar analogy for a carbon dioxide assimilating cell, we must imagine an enterprise which has specialized in the production of hydrogen and oxygen through electrolysis of water, the hydrogen being used, for example, for the hydrogenation of carbon dioxide to methanol and hydrocarbons. Part of the energy intake—electrical power in this case—is used for the synthesis of more complex substances from the hydrocarbons such as different chemical and plastic products, some of which are then used to develop the business, and so on.

We can carry these analogies to absurd lengths, but only for fun. Let us suppose that a business which specializes in gasoline refining, for some reason that is known only to a few, decides to collaborate with a company which specializes in carbon dioxide hydrogenation and, instead of utilizing natural rock-oil changes, uses the synthetic hydrocarbons of this second firm. The carbon dioxide produced is promptly delivered for further hydrogenation into hydrocarbons, etc., etc., and the companies combine into a fully occupied unit with all their products circulating within their own organizations, at the cost of a considerable amount of energy. The economic basis of such an arrangement is, of course, quite absurd, as well as the motive behind it, but such a system has a certain ironic analogy with the system by which carbon dioxide is produced and assimilated on our earth. In this case, the energy in-

---

actions. The hydrogen ions are transformed into enzyme-bound hydrogen which reacts with carbon dioxide as in systems *a* and *b*. In this process iron salts at the Fe^{3+} level are formed. A limited number of cell types belong to this group, which as a whole is designated iron bacteria. (*d*) A regular, food-consuming type of cell: the counterpart of systems *a*, *b*, and *c*. Hydrogen is split off via enzymatic reactions coupled in series and eventually combines with oxygen in the formation of water; this is an energy-producing process which is partly utilized for the formation of new cell material via active phosphate units and certain transformed nutritional material. The latter is represented by other cells from all the groups *a*–*d*. Carbon dioxide is released from the system and is utilized by *a*, *b*, and *c*. The co-operation between *a* and *d* represents life on earth.

take—in the form of sunlight—is free and this gives it a sounder economic basis.

Life on our earth, with all its enigmatic lack of motive, is therefore a system within which two opposing tendencies made up of innumerable cell systems and their positive and negative contributions operate. Bearing this in mind we can now consider the question of the ways in which life on other planets may differ from our prototype.

In doing so we must differentiate between the composition of the material and the form of the energy-producing systems. To begin with, I should point out that there are many who are happy to accept almost everything within the carbon compound group. I personally think that this is the wrong starting point for an analysis. The components of the cell systems on earth——carbohydrates, fatty acid derivatives, nucleic acids, and above all, proteins—need not be classified as local specialties of biochemical synthetic activity. They should be regarded instead as local expressions of a universal and general tendency exhibited by carbon material in its organization of itself into something living. If we think about the problem we shall find that the chemical starting point for any planet capable of developing life is the same all over the universe and that the lines of development must follow certain common channels.

In the first place, we should bear in mind that all stellar or interstellar material which has been cooled while aggregating into planets has, by and large, the same chemical composition. Any naturally occurring planet components beyond our known series of elements, with lead-208 as the heaviest stable component, are quite out of the question, even if some enterprising chemists are sometimes able to increase the series experimentally from uranium-238 via plutonium and other intermediate products up to advanced elements with a pronounced instability. Our own record is nobelium-252 with a half-life time of approximately ten minutes. These experiments apart, the starting-point for all planets some five billions of years ago must have been monotonously the same: the result of the cooling down of a material which had oxygen, silicium, and iron as its main elements at the first aggregation, with cover of hydrogen. Our own planet can in all probability serve as a model for universal planet formation from a chemical point of view. The variation between very hot and very cold and small and large planets is, from the chemist's point of view, to be expected as a general phenomenon within the cosmos that we can survey.

The development of life from accumulated inorganic material and simple compounds of carbon, hydrogen, oxygen, nitrogen, sulfur, and phosphorus—to select some safe components—is in its simplest form a dynamic period of transformation initiated by a temperature difference. Such a development can take place, in its simplest form, in the chemical turbulence between a hot planet surface initiating disintegration and reactivity and a relatively cool atmosphere that allows a recombination of reactive units. From a general chemical point of view carbon compounds show the most variable component material. During successive cooling most inorganic components will stabilize into definite compounds which will certainly undergo some transformation in the surface layers of the planet, but they do not have the capacity to develop in the complicated ways that carbon compounds develop.

Later phases of a planet's existence will be characterized by the developmental pattern of its carbon compounds, according to certain limiting factors. The salient point is that of temperature. Cooling much below 273° K = 0° C causes a successive decrease in the reactivity of the carbon compounds. The cycle begins to slow down, the chemical equilibria become icebound, the thermically active material is locked up in chemical sleep. This probably represents the state of affairs on Jupiter and the giant planets, where what was once active gradually ceased to be active, where primitive life—without distinct organisms—in thermic agitation is now deep-frozen under great pressure, and where the atmospheric changes that can now be observed represent an echo of what was once a promising development but which is now at an end.

At temperatures above 100° C–200° C, however, the chances that more complicated carbon compounds can be formed become slim. The material is exposed to such destructive thermic forces that the more advanced molecular structures can hardly exist for any length of time. If we look at the diagrams in Figures *6a* and *6b* we can see how narrow the temperature range is within which carbon compounds can combine so that they can be realized as active life.

The earth's position and temperature conditions have been and still are favorable for the development of carbon compound reac-

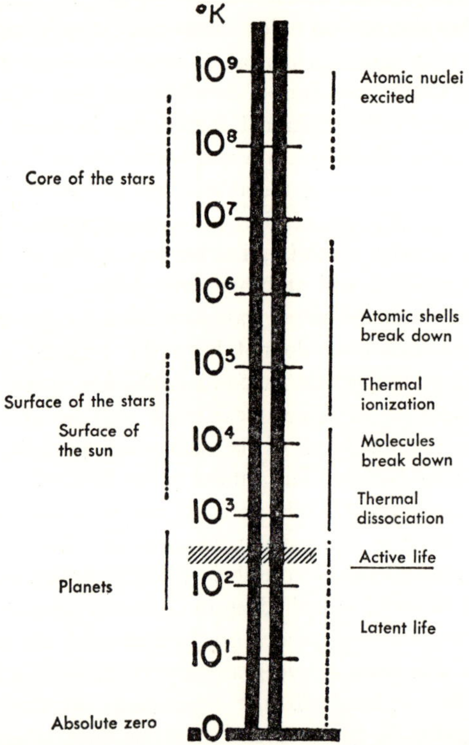

FIG. *6a*.—A diagram showing the temperature conditions in the universe. It illustrates how narrow the range is in which life as we know it on earth can exist. The scale is in absolute temperatures with 0° K at absolute zero and 273° K at 0° C. What has been described as latent life represents organism systems capable of surviving temporarily in a temperature interval below or above what has been indicated as optimal for an active, developing form of life at a cellular level.

tions. When we consider the role that water plays in the reaction patterns of organic compounds, we see that this gives us a point of reference, mainly the point 0° C, for the development of advanced life processes. Carbon compounds as a whole need special activating agents to display some of their reactivity under this temperature. Over 0° C water can serve as a medium for a number of reaction processes both for low-molecular as well as high-molecular components. In particular, it permits certain catalytic features of high-molecular components to develop fully.

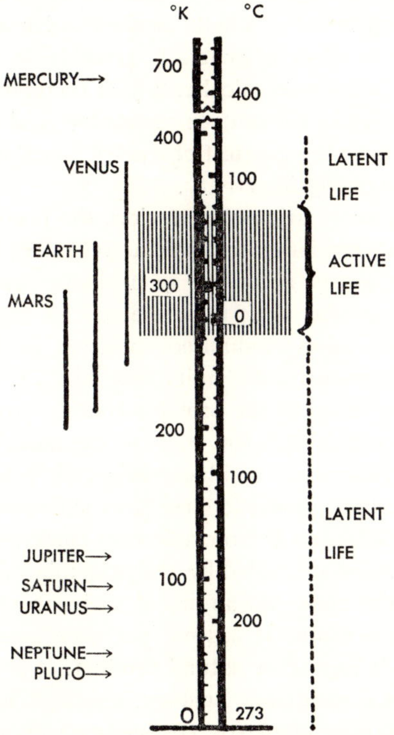

FIG. 6*b*.—A diagram showing the temperature conditions on earth and on the other planets within our solar system. It can be seen that only the earth, Mars, and Venus are able to sustain active life in the form that it has developed on earth, that is, organic material of a high degree of complexity with cells and multicellular organisms.

Keeping this in mind, we arrive at the concept of heterogenous catalysis via high-molecular systems. The catalytic units of our cells are without exception proteins, the enzymatic activity of which is intimately connected with water both as a disperser and as a solvate. This leads us to the question of the universal nature of biocatalysts.

What we know about organic chemistry and the reaction possibilities of carbon compounds is universal in its application, irrespective of the locality. The one and a half million compounds that we know of so far, some of which are natural products but the majority of which are synthetic products, give us a fairly complete knowledge of what is generally possible in this field. Our statement about organic-chemical catalysts therefore has a certain authority for as far as we can see the amino acid condensates of the protein type have the greatest chance of functioning as carriers of high-molecular catalytic activity.

From a purely pragmatic point of view, the protein structure is the most favorable carrier of catalytic activity. We have a number of units joined together according to a similar fundamental pattern into high-molecular structures, the specific three-dimensional structure of which allows unlimited variation in the patterns of association with low-molecular units (see Fig. 7).

The amino acid units of the proteins contain groups designated as "side chains" which represent most of the basic organic-chemical structures: hydrocarbon residues, carboxylic acid residues, simple and cyclic amines, aromatic structures with simple phenyl arrangements, phenols, and heterocyclic ring configurations. Protein structure is, on the whole, ruled by the principle: simplicity in the combination of the units, variability in their structure, versatility in the pattern as a whole. It has not yet been possible to trace a similar principle in any other natural product. It is extremely difficult to give even a reasonable analogy, although it might be possible to invent a type of plastic condensate, copolymerized from some twenty different units carrying the same chemical characteristics as the known amino acids.

The more one considers the problem, the more it becomes apparent—from both the organic-chemical and biochemical point of view—that proteins of all kinds represent the most plausible results of the transformation of carbon compounds in a milieu where they have been undisturbed in the presence of water for a couple of billions of years. It would be too complicated to present all the arguments for this theory, for they are of an advanced kind, but I can give one detail here, that is, the experimental method of simulating the formation of amino acids in a prebiotic time.

Taking the classical experiments of S. L. I. Miller, who in 1953 produced traces of amino acids by means of electric discharges in a gas mixture containing methane and ammonia as a basis, many research groups in the United States have studied the formation of amino acids from extremely simple materials. To judge from the publications that have appeared so far, the results indicate that the natural tendency for amino acids to form from any kind of carbon material under the influence of ultraviolet radiation and/or electric discharges is unexpectedly pronounced. This can and does occur in the upper strata of our atmosphere, but on a small scale now in comparison with billions of years ago.

From this point of view, amino acids become a universal factor indicating the universal existence of proteins. This, therefore, also applies from a chemical point of view even to enzymes and with greater probability to the organization of cells and organisms, insofar as they have universally common features.

The same is true for carbohydrates. They can be thought of as almost inevitable products of simple carbon compounds in the presence of water. Fatty acids of all kinds can, of course, be produced by the simplest oxidation processes of prebiotic hydrocarbons. Every kind of organic-chemical analysis presents the same conclusion of the universal transformation possibilities of primary carbon material that can be stated as the *inevitability of the formation of certain groups of substances* belonging to present-day terrestrial chemistry.

These arguments naturally seem dogmatic to those who want to think that life, even intelligent life, could exist in forms quite different from those on earth. Why cannot there be other high-molecular systems than proteins or other systems than amino acid condensates? Could not carbohydrates act as carriers of catalytic

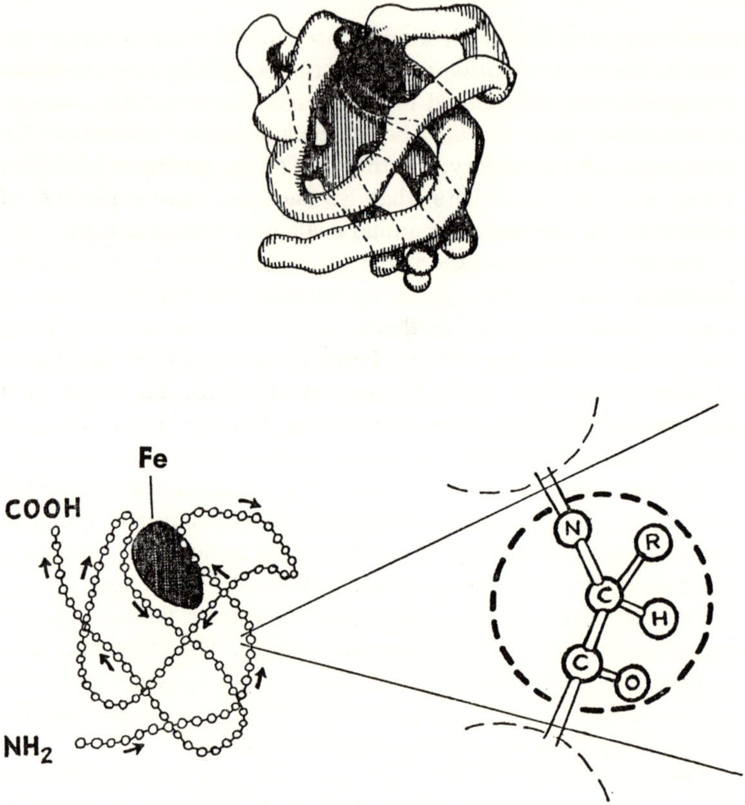

FIG. 7.—A schematic representation of the structure of a protein from three different aspects, *a, b,* and *c.* Diagram *a* (*top*) shows the protein as a longish unit coiled up according to a strict pattern into the specifically determined tertiary structure of the protein. The long unit, a peptide, has an amino group, —NH_2, in one end and a carboxyl group, —COOH, in the other. It encloses in a definite way a built-in structure with a central iron atom, Fe; the latter arrangement is one of the many existing variations on the same theme. As indicated in diagram *b* (*bottom left*) and *c* (*bottom right*), the peptide chain is composed of a large number of units, in this case 153 amino acids. Every amino acid has the same general formula: NH—CHR—CO—, in which R is an atomic structure with about 15 variants, each of which is characteristic of every amino acid. The sequence of the different types of amino acids within the peptide constitutes its characteristic secondary structure. Many

properties? Why must water be the solvatization agent; why not ammonia or ethyl alcohol? Why cannot there be radioactive sources of power instead of sunlight and the conventional supplies of hydrogen sulfide, Fe^{2+}, etc., etc.? Why should there be cells or metabolisms at all? Why should there not be something quite new?

Why not?

The flaw in such an argument—stimulating though it is—is the assumption that it is impossible for us, in our part of the universe, to work out the development of life in other parts of the universe from a chemical point of view. But that is exactly what we can do. The chemical laws for the combination of substances are independent of time, space, and locality. What we are trying to find out is the most probable state of chemical activity, the most probable combinations of carbon compounds given the same initial position at a given time. Both on earth and elsewhere.

By following this reasoning we end up with a system—irrespective of our preconceived ideas—which is strikingly analogous to that of life as we know it on our earth. Many alternatives have been suggested and one particular idea has been developed in science fiction stories, that is, that silicon compounds might function as substitutes for the conventional carbon compounds. But all the experts on silicon chemistry tell us, with regret, that silicon polymers of the protein type are in any form always rather labile, either in water, in liquid ammonia, or in ethyl alcohol. Complex carbohydrates which have about twenty components have little chance of binding catalytically active groups into catalytic units of the enzyme type. Reactions often take place quite easily in

---

proteins contain more than one peptide chain, and the numbers of amino acid units may in certain cases amount to several thousand. The determination of the sequence of the amino acids within a peptide is an extremely difficult task, and it has only been carried out for a few simple cases. The determination of the tertiary structure has only recently been carried out in two cases. The schematic structure in this diagram represents the protein myoglobin, the structure of which has been determined by X-ray analysis by an English research team after ten years' work. (From a drawing in *Nature,* CLXXXV [1960], 426.)

liquid ammonia, but it would be unlikely to find a system of cyclic reactions which does not require metallic sodium or other rare stimulants. That life can exist in pure ethyl alcohol is an interesting but not very promising theory in view of the lack of certain stimulating metals which could hardly be formed in the free state during any geologic processes. The theory that radioactive disintegration could act as an energy source could only be accepted if the disintegration processes had been continuing for at least one billion years. Otherwise life would be hectically short. The following comparison is quite revealing. The amount of radiation energy received by the earth in one minute is 5×10^{18} cal, which is equivalent to the amount of energy produced in the same time interval by 10^{18} tons of uranium, half-life time 4.5 billion years. It is hardly necessary to say any more.

As for energy sources it would seem unnecessarily ambitious to use a system that did not involve directly the radiation of the sun or the chemical divergencies of the oxidation level which exist in the outer envelope of the planet as a heritage from the first aggregation, for example, iron at the $+2$ and $+3$ level. The theory that I personally find most plausible for the development of life on earth is based on utilization of the energy differences between hydrogen sulfide and sulfur and between Fe^{2+} and Fe^{3+} during the first three billion years. Solar energy comes into the picture only about one billion years before *our* time, in connection with the development of the first photosynthesizing organisms. This development is possible throughout the universe, and it is the one which is the most likely to occur and the easiest to execute.

This extremely short survey of the characteristic properties of living systems is of course highly naïve, but a fundamental analysis would take volumes and volumes. The basic points I have been making about living systems in general are the following.

1. They are based on cyclical conversion processes with carbon compounds as components.

2. In an advanced stage they have systems of high-molecular-weight biocatalysts with amino acids as the dominating structural elements which are analogous to our proteins.

3. They utilize chemical energy differences as propelling forces;

in advanced stages sun energy and photolysis of water are fundamental factors.

But the ways in which these systems develop according to the principles given above is quite another matter. It is by no means certain that organisms, as *we* know them, are a prerequisite of life. A co-ordinating system of small lakes covering a planet's surface, with half of the lakes engaged in the catalytic activation of carbon dioxide into complex organic material and the other half catalytically decomposing the products of the activity of the former units into carbon dioxide, and a sun-activated flow of water around the planet, would undeniably constitute life, but it would hardly correspond to our idea of organism life. If we should meet something like this on our future travels, we would hardly recognize at first that we were dealing with something living, for we would not see any mass, body, or anything moving, but only a global activity in chemical serenity. We would find something that only exists and lives. It would be difficult, too, for us to comprehend life in an atmospheric envelope as well as collective life on a global scale, with water particles and the components of fog and perhaps clouds forming microlocalities of enzymatic transformations of carbon compounds according to the general scheme that is valid for all forms of life. On Venus especially there is reason to keep a lookout for such systems—systems which have nothing to do with awareness as such but which constitute life without any of its higher and more complex manifestations.

And so we must face the question of whether there is life on Venus and Mars, and what its forms may be. It would probably be safer to say, "let us wait and see." One cannot resist, however, with all the recklessness of a gambler taking a chance on unseen cards, if one has a certain know-how about the game. So let us make an extrapolation.

A preliminary analysis of the conditions on Venus tells us that its present climate hardly permits any form of living structures at the ground level, since the temperatures, as we observe them, probably exceed what proteins can stand without decomposition. So it would be safe to conclude that there cannot be any form of life at the ground level of the planet. But we cannot discount the possibility of there being life in the upper strata of the atmosphere, as an airborne system of active, co-operating chemical units exchanging chemical substances in temperature ranges of about 0° to 50° C. Any such organisms would probably not be very large. The largest airborne unit that *our* form of life has produced is, in the insect world, the extinct giant dragon-fly *Meganeura,* which flourished during the carbonic era and which was one meter wide. We have also produced overgrown pterodactyls. The now extinct Pteranodon is the best example of this group; it had a wingspan of 5–8 m, but probably it was a poor flier. The most advanced airborne units at present are the South American condor and the subantarctic albatross. They are both well-developed birds with a good aerodynamic construction. All these are respiratory organisms and therefore dependent either directly or indirectly on the flora of the earth. If one does guess about the condition on Venus on the basis of the theory of completely airborne life, it would be best to assume a collaboration between something like aeroplankton, photosynthetics, and saprophytes and to discount altogether any higher forms of development. It is probably wiser not to speculate about the existence of any self-propelled units, for such units would have had to have undergone a long period of evolution on the ground, which is unlikely because of the ground temperatures. It could be argued that the deterioration in the climate in the form of increased temperatures occurred at such a late stage that creatures capable of flying were already present and only adapted themselves to the new state of affairs by living on aeroplankton whirled up to the upper strata and by reproducing quickly so that they do not have to descend to the ground regions. Bearing all this in mind, we nevertheless have difficulty believing in the theory that there are intelligent, compact inhabitants on Venus who are responsible for flying saucers, etc.

A completely airborne cell organization might, however, achieve such a correspondence between its units that it could develop a certain group awareness (see chap. xi). The question is whether such a dispersed form of life could ever develop the heavy industry that is essential before communications with neighboring planets can be achieved. If we ever encounter any form of life on Venus, it will probably be from the perspective of the microscope.

Mars offers more possibilities. The thermic range of the planet is such that proteins and water in liquid form can exist in temperatures which periodically may rise as high as +20° C. Judging from our own conditions, what we should find on Mars should be something that could survive in latitude 75° N. and 65° S. on our earth under conditions involving an excessively low atmospheric pressure and a pronounced lack of water. In a more detailed discussion of this problem in chapter vi, I conclude that it will probably correspond to the lichens of arctic and antarctic types. But Mars has had the same possibilities as the earth over a long period to develop self-propelled organisms. It is true that partial anaerobic conditions have, from our point of view, not produced any higher organisms than our present tapeworm, but the situation on Mars two billion years ago was certainly more favorable as far as oxygen supply and water content are concerned. If we find any organisms on Mars we may conjecture that they are recessive forms of a past flora and fauna as they were during our Cambric era. They have survived, but they have not had a suitable external environment in which to develop into more advanced forms of life.

It is difficult for us to understand that there is only a very narrow range of temperature and gravitational and chemical conditions in which something that lives can develop into something that lives *consciously*. Life in different forms probably occurs in all parts of the universe, both Venus and Mars will certainly have (or have had) some forms of life, with many characteristics that seem to be common both to them and to us. On the other hand, the chance of finding some degree of awareness, someone or something with which to communicate, is small. At least on the neighboring planets.

Even if we encounter a form of local life on one of the neighboring planets, it is by no means certain that we would recognize an awareness in what we see around us even if it is there. We shall have to acquire a sensitivity toward the nuances in the behavioral pattern of a strange organism world which may indicate that the whole is something more than the seemingly simple units that we come across. Our training for the journey will have to include the development both of our powers of objective observation and of sensitivity. In this way we will learn not to expect things that may not exist at all and not to allow subtle nuances which may indicate something living to pass us by.

CHAPTER V

VENUS: THE VEILED UNKNOWN

By the time manned expeditions are made to Mars and Venus, modern astronomic techniques should have provided so much preliminary information about the surface conditions of these planets that the first explorers should be well prepared for the peculiarities of the local milieu. Exploration of the moon will of course be the preliminary phase in which the equipment and the material to be used for the construction of the first outpost on Mars and Venus will have been tested. A number of unmanned observation probes will also have been sent out to, around, and back from both Mars and Venus to register and transmit data from an observation at close quarters—a project which is already within the realm of possibility.

Although Venus is nearer to the earth than Mars—it is at minimum only some seventy-five million kilometers away—a preliminary analysis of local conditions presents certain additional difficulties because of the dense and non-transparent atmosphere surrounding the planet. The data we have accumulated by distant observation from the earth are reliable in certain respects, but the composition of the atmosphere, its temperature and pressure conditions, and the structure of the ground surface are problems which are still under discussion and no general picture has yet been woven out of the data we possess.

Table 5 summarizes what we know definitely about Venus in

comparison with the earth. These data by themselves only tell us that the Venus year is approximately two-thirds of an earth year and the gravity on the surface of Venus should be slightly less than that on earth. The gravitational acceleration for Venus has been calculated as 860 cm/sec^2 as compared with 981 cm/sec^2 for the earth so it would not give us any extra advantages on long marches, as on Mars. As far as temperature is concerned—from a comparison between the solar constants of 3.2 and 1.9 respectively—it is clear that Venus must on the average be warmer than

TABLE 5

Data	Absolute Values		Relative Values (Earth = 1)	
	Venus	Earth	Venus	Earth
Mass	4.900×10^{27} g	5.975×10^{27} g	0.82	1.00
Radius	6200 km	6371 km	0.97*	1.00*
Density	5.0 g/cm^3	5.5 g/cm^3	0.91	1.00
Orbital radius	108.1×10^6 km	149.1×10^6 km	0.72*	1.00*
Revolution period	225 days	365.256 days	0.62	1.00
Solar constant	3.2 cal/cm^2/min	1.9 cal/cm^2/min		
Rotation period	Unknown; between 15 and 60 days	1 day		

* Mean values.

the earth. The question is, How much warmer? The composition and pressure of the atmosphere and the rotation pattern of the planet are also problems which are intricately interwoven with each other. Their solution is essential before we can work out our chances of surviving a landing, building a base, and exploring a part of the planet's surface.

Our knowledge of the remarkable density of the atmosphere on Venus dates back to 1761 when T. Bergman at the Uppsala observatory in Sweden happened to observe a light ring around the planet on one of the rare occasions when the planet was exactly between the earth and the sun and was thus sharply outlined as a small black dot against the sun's disk. On the other two occasions when

Venus moved into this position in 1769 and 1886—it is a highly infrequent event; the next time it will happen will be in 2004 and 2012—the phenomenon was confirmed and explained as the refraction of sunlight in a dense atmosphere around the planet. Attempts to determine the composition of this atmosphere by the use of a spectroscope yielded few results for a long time. It was not until 1932 that Adams and Dunham at the Mount Wilson Observatory in the United States discovered the presence of certain bands in the absorption spectrum of the planet which were derived from carbon dioxide, CO_2.

The discovery of carbon dioxide in the atmosphere of Venus, was, in many ways, a complete surprise, especially since the data revealed a considerable amount. To compare it with the earth's atmosphere, we can state that carbon dioxide in our atmosphere, calculated per cm^2 of the earth's surface, represents a 2.2 m high column. According to the measurements of Adams and Dunham, the corresponding measurement for carbon dioxide in Venus' atmosphere is approximately 1000 m—if the calculations are made of the gas at normal pressure and 0°—that is, 450 times the value for the total amount of atmospheric carbon dioxide on earth. We must also bear in mind the fact that the measurement of Venus' atmosphere is based on the passage of light through the upper, transparent layers of the atmosphere, above the covering clouds, so that the absolute figure for the carbon dioxide content of Venus' atmosphere must be even higher.

An approximate calculation shows that the actual pressure in an atmospheric region where a 1000 m value for carbon dioxide content is registered, corresponds to 0.14 atmospheres. The question is, however, At what height is this region located in Venus' atmosphere? It is clear that it is above the clouds, but at what height? Fifty km? Thirty km? Fifteen km? The pressure in Venus' atmosphere should decrease by a factor of 2 every 5.5 km; so counting from ground surface, we can make the suggestions given in Table 6.

Some investigators agree that the mean height of the cloud covering has a value of approximately 30 km, which would mean a ground pressure of carbon dioxide of 5 to 10 atm. But the value of 0.14 atm in the measured region may be too high, which would mean a considerably lower ground pressure. The data found in literature on the subject diverge, however, and extreme values which have been published as alternatives for the ground pressure range from 1 to 2000. The Americans Shaw and Bobrownikoff have suggested an approximate value of 100 atm, a value which is dubious from a purely chemical point of view. As our start-

TABLE 6

THE CO_2 PRESSURE AT THE GROUND SURFACE OF VENUS AS A FUNCTION OF THE HEIGHT OF THE 0.14 ATM REGION

Height	50	44	38.5	33	27.5	22	16.5	km
Pressure at the ground surface	80	40	20	10	5	2.5	1.25	atm

TABLE 7

Carbon Material	Calculated as 10^{12} Tons of Carbon	Calculated as 10^{12} Tons of Carbon Dioxide	Calculated as Kg of CO_2 per Cm^2 of Ground Surface (In Atm)
Atmospheric carbon dioxide	0.6	2.4	0.00044
Carbonates and carbon dioxide in solution in the seas	50.0	190.0	0.03
Carbonates in the earth crust	12,000.0	44,000.0	8.5*

* Possible maximum value: 10 atm.

ing point we can use the careful inventory of the carbon supplies of the earth made by V. M. Goldschmidt, K. Rankama, and other geochemists. According to the present distribution of carbon between earthcrust, sea, and atmosphere we can calculate from Table 7 the pressure that would exist if all this carbon occurred as carbon dioxide in the atmosphere. As we can see from this table, if all the carbon was oxidized and gasified as carbon dioxide in the atmosphere around our planet, it would result in an average pressure of 8–9–10 atm. This would be the absolute maximum for our planet! How about Venus?

Everything seems to favor the assumption that the total amount of carbon on Venus cannot be greater but is rather less than on earth. The carbon that accumulated five billion years ago on and inside growing planets was most probably in the form of methane and other hydrocarbons, and the enormous accumulations of methane on the giant planets give unequivocal evidence for this hypothesis. With this in view, it seems highly unlikely that a considerably larger amount of hydrocarbons should have been allotted to Venus than to the earth in the small area which makes up the Venus and earth orbits. Probably it is the other way around, with Venus having a smaller share of hydrocarbons than the earth, which in turn has a smaller share than Mars, and so on to an increasing degree, with a maximum somewhere in the Jupiter-Saturn region. We must also take into account the fact that Venus for the past five billion years has had a warmer climate because of its position closer to the sun. As a result there must have been much greater evaporation of volatile substances on Venus. From this point of view, CO_2 pressure of approximately 5 atm could be considered as a maximum value, in which case *all* the available carbon on the planet would have to be concentrated in the atmosphere with nothing in the bedrock. This seems unlikely, but perhaps it is worth discussing in a little more detail.

Let us assume that the planet has had such a high surface temperature during a certain period that the chemical reaction

$$R{-}CO_3 + SiO_2 \rightarrow R{-}SiO_3 + CO_2$$

has gone in the direction of the arrow and that at the same time the main part of the water on the planet has been lost through photolysis and loss of hydrogen. It could then be possible to arrive at a state where silicates dominate in the bedrock and carbon dioxide in the atmosphere. This same state cooled down from a chemical reaction temperature of approximately 600° C to, let us say, 200° C or lower will in the absence of water become stationary for long periods of time. The reaction

$$R{-}SiO_3 + CO_2 \rightarrow R{-}CO_3 + SiO_2$$

is favored by low temperatures and a high water content in the environment, which does not seem to be the case on Venus. The main point of this whole discussion is whether all the carbon on Venus can occur as carbon dioxide. If this is so, the water content must be low. If all the carbon is present in the atmosphere as carbon dioxide, then the ground pressure cannot exceed 10 earth atm, corresponding to 8.8 Venus atm. The value is certainly smaller.

Now that we have got so far it is worthwhile asking how much we know about the occurrence of other elements in the atmosphere of Venus. We have very few observation data on which to build, but we can make some assumptions with a certain amount of confidence.

Nitrogen should occur on Venus as it does on earth and in the same amounts. This assumption is based on the same arguments as in the case of carbon and carbon dioxide. At the same time when the planets were formed, nitrogen existed mainly in the form of ammonia, NH_3, and to some extent as the compounds HCN and $(CN)_2$. In contact with hot metal oxides, the nitrogen compounds were transformed quite rapidly into molecular nitrogen, N_2, according to the formulas

$$2NH_3 + 3MeO \rightarrow N_2 + 3H_2O + 3Me$$

$$2HCN + 5MeO \rightarrow N_2 + H_2O + 5Me + 2CO_2 .$$

The giant planets provide evidence of the nature of the original nitrogenous material, since both Jupiter and Saturn have large amounts of frozen ammonia, NH_3, in their atmospheres.

The reactions symbolized above must have worked almost completely on earth, where more than 97 per cent of all nitrogen is in the form of gaseous N_2 in the atmosphere, with a pressure of 0.8 atm. It is reasonable to assume a similar course of events on Venus, the CO_2 atmosphere of which ought to have a partial pressure of N_2 of approximately 0.5 atm.

The question of whether water occurs on Venus is another problem. Earlier investigations of the spectrum of Venus' atmosphere revealed that a maximum of 5 per cent of the water content of the earth's atmosphere could occur on Venus. Recently some observations have been made by means of stratospheric balloons equipped with powerful telescopes. Some spectra from Venus, obtained above the main part of the earth's atmosphere at a height of 20,000 m, suggest the presence of water to the high degree of 10 per cent of the corresponding content in the earth's atmosphere. It is not a high percentage compared with that of carbon dioxide and nitrogen, but it is enough to be of significance for any possible explorers of the planet. Recent observations (1964) suggest that the clouds themselves are composed mainly of ice crystals suspended in the dense atmosphere.

TABLE 8

Atmospheric Components	Venus Atm	Earth Atm
Nitrogen	0.5–1	0.8
Oxygen	....	0.19
Argon	0.01	0.01
Carbon dioxide	1–10	0.00044
Water vapor	0.001	0.005
Total pressure	1→10	1.0

Another atmospheric element undoubtedly present on Venus is argon, in the form of the argon isotope ^{40}A, which is formed by the radioactive disintegration of the potassium isotope ^{40}K. Potassium must be present in the bedrock of Venus, and since the velocity of radioactive disintegration is independent of time, location, and exterior circumstances, the atmosphere on Venus must contain a percentage of argon, which, analogous to the conditions on earth, would add 0.01 atm to the total pressure of the atmosphere.

A preliminary analysis of the probable composition of the atmosphere on Venus compared with that of the earth is shown in Table 8, in which the uncertain values, of course, are those for carbon dioxide and water. It is probable that the upper boundary of the cloud-covering is between 25 to 35 km from Venus' surface. The sky above is, if not blue, bluish black. An arriving expedition circling around the planet a couple of times would have the chance to study the reflection capacity of the cloud covering in the light of an enlarged sun, the disk of which will seem at this distance to be about 1.5 times larger than from the home planet. The reflection capacity of an astronomic object, the so-called albedo value, can have a maximum value of 1. For Venus it is somewhere between 0.59 and 0.76—an unusually high value. The albedo of the moon by comparison is 0.07 and that of new-fallen snow 0.78.

A great many different suggestions concerning the Venus cloud covering have been put forward over the years, from dispersed water drops or ice crystals—that is, conventional elements of our own clouds—to microscopic dust particles like sand or different organic compounds. Probably the main components are ice crystals with the addition of dust particles from the lower turbulent atmospheric strata. There is a close connection between this problem and that of the atmosphere and ground surface, which in the last few years has been investigated by means of short-wave techniques.

Until 1956, temperature measurements were based on the infrared radiation emitted from the upper strata of the atmosphere, that is, from regions above the cloud covering. The values obtained ranged from —35° C to —45° C, the light side being a little warmer than the dark one. Further measurements by this method are however impossible, for the infrared cannot penetrate the cloud covering. Measurements by ultrashort-wave techniques—microwave measurements in the 3 cm region—have enabled us to penetrate the cloud covering considerably farther and have provided data about the temperature at or near ground level. The results are surprising. All the measurements indicate temperatures of between 200° C and 400° C, with a clear difference between day and night temperatures.

There is one extraordinarily interesting phenomenon peculiar to Venus which cannot be explained solely by its closeness to the sun. A comparison of the average temperatures of Mars and the earth with their respective solar constants of 0.97 and 1.92 would lead us to expect that Venus with its solar constant of 3.8 would have an average temperature of under the 100° C mark (see Table 9). If we can rely on the ultrashort-wave measurements to give us a true picture of the ground temperature and not of some of the inner atmospheric strata, Venus' atmosphere, with a high thermic emission in the 3 cm band, must be an extremely effective insulator for heat radiation in long-wave infrared. It cannot be the result of carbon dioxide alone; there must be something else that we have not yet grasped.

About the particles in the atmosphere, the experts favor the view that the high temperature difference between the ground and the upper strata of the atmosphere must lead to pronounced turbulent currents, some of which must exist as strong winds at ground level. "Wind" is probably too mild a term to describe the atmospheric disturbance at the ground level. A great deal of loose gravel and dust must be whipped up from the ground surface into the upper

TABLE 9

	Solar Constant Cal/Cm²/Min	Mean Temperature Calculated	Mean Temperature Measured
Mars	0.97	−15°	− 15°
Earth	1.92	+ 5°	+ 5°
Venus	3.82	+50°	>+200°

strata; this would provide a relatively simple explanation of the opaque cloud structures of the atmosphere: dust in lower strata, ice crystals in higher levels. On the hot ground surface, sand storms rage over the plains. During the day there are intervals of semi-darkness; during the night it is quite pitch-black.

Any expedition circling the planet on a couple of orbits and preparing to land would be prepared for all this and for the contrast between the reflections of the intense sunlight in the upper layers of the cloud cover and the dark hot activity at ground level. All the radar and microwave equipment is turned on; the ship starts its braking maneuvers, plunging eccentrically down into the denser layers of the atmosphere in order to reduce speed for the final drop into the dark unknown. The size and mass of the planet do not differ greatly from those of earth, and so these maneuvers can be practiced at home. The final phase of the landing will probably be a vertical descent similar to the landings on the moon, but with radar eyes of all kinds peering into the dusky darkness. Registering reflexes against bare ground, sand, hills, more sand. . . .

After the first ground contact is made—probably a difficult feat in the gusty and stormy weather—the expedition members will have an opportunity to inspect the landing ground for the first time, so far as they are able to see. After a good deal of forethought the landing place would probably have been chosen in a relatively cool region and in an area having some light during the morning period so that at least the immediate surroundings could be seen.

Those members of the crew who—no doubt somewhat hesitantly—climb out onto the side of the ship in their heavy pressure- and heat-insulated space clothes will probably make some sarcastic remarks about the mild weather and the marked activity of these regions. The temperature is unbearably high and prohibits any further excursions. The pressure is a couple of atmospheres higher than the one we are used to on earth. Heavy clouds rush across the sky toward the south; the light changes from a yellowish-gray mist to that of a northern November morning at its gloomiest. It makes one think of the approaching period of about fourteen days of darkness, when not a thing will be visible in the sandy darkness, full of turbulence, and one begins to wonder whether one will be able to stand this semidusk with any degree of equanimity. The planet leaves a great deal to be desired from the point of view of comfort. What possibilities are there for human colonization? What chances have there been in these areas for the evolution of some form of indigenous life? Has life ever existed here, or does it still exist in some sheltered corner? If so, where?

Venus has some inescapable disadvantages as far as the construction of any kind of observation base is concerned. Any base would have to be well insulated from both heat and pressure, and refrigeration equipment would be a necessity. For any kind of outdoor work a protective dress resistant to changes both in pressure and temperature would be unavoidable, and microwave and infrared equipment would be needed to extend the scope of vision in the dusk. Energy to provide electricity for the base on Venus could probably be provided by windmills; that is at least one point in its favor. Chemically speaking, the surplus of carbon dioxide in the environment would seem to be worth preserving as a basic material for the cultivation of plants, provided that water can be found and energy can be produced on the spot.

The question of whether plants can be grown on Venus can be considered as solved now that we know that water *de facto* exists in the atmosphere to a limited extent. The problem will be an entirely technical one, that is, the separation of H_2O, CO_2, N_2, and A by distillation at low temperatures—something that is done at every carbon dioxide factory on the earth. Thus the conditions necessary for the successful greenhouse cultivation of plants with the immediate aim of manufacturing local oxygen can be created on Venus in the same way as on the moon. The process could be carried out according to a technical variant of the theme

$$\underset{\text{carbon dioxide}}{CO_2} + \underset{\text{water}}{H_2O} \xrightarrow{\text{light energy}} \underset{\text{plant material}}{CH_2O} + \underset{\text{oxygen}}{O_2} \; .$$

From a practical point of view the plants chosen should be able to thrive at the highest possible temperature, be economical of water, grow rapidly, and be useful in other ways. As far as light sources are concerned, it seems better not to depend on the sparse light from the outside but on energy supplied from the windmills. In other words the vital parts of the base should be underground at least for the time being. For extra supplies of oxygen to the base, a plant for the electrolysis of water could be established, the hydrogen being released into the open for the time being. A compact nuclear energy unit will of course be necessary to the base during the first stages of the establishment of the settlement.

As the construction of the base proceeds there will be time every once in a while to speculate on the fundamental probem of why the external milieu on Venus has turned out to be so extremely dissimilar to that of the earth. The density and mass of the planet, both of which correspond very closely to those of the earth, suggest that the material from which both planets were formed was essentially the same. In both cases we have to imagine that the planets, after their first phase of growth from cosmic gravel, were made up of warm central bodies of metal oxide-silicate type surrounded by dense atmospheres of hydrogen and hydrated compounds in general: hydrocarbons, cyan compounds, water, ammonia, and later carbon monoxide and carbon dioxide. The reaction between the hot surface and the atmospheric components gradually resulted in a local oxidation with the participation of metal oxides. After a time the atmospheres came to be dominated by water, nitrogen, and carbon dioxide, with smaller amounts of different kinds of organic compounds which on the earth gradually combined to form polymeric material with catalytic properties, the forerunners of life on the earth. How did the evolution on Venus differ from that of the earth?

When we think about this problem we should of course bear in mind the positions of the two planets in their respective orbits around the sun. Venus' position—about forty-five million km nearer the sun—made it inevitable that the cooling processes would take place much more slowly than on the earth and, moreover, that the double radiation received from the sun would have a fundamental influence on photochemical processes of all kinds in the atmosphere.

The first question to be considered in this connection is that of photolytic decomposition of water via ultraviolet radiation in the upper strata of the atmosphere. This reaction, which results in the production of hydrogen and oxygen, gives the hydrogen, with its low molecular weight, a chance to escape from the planet. The critical velocity necessary for a departure from the exterior atmospheric layers of Venus into the interplanetary vacuum is 10.2 km/sec, which is somewhat less than the corresponding figure for the earth: 11.4 km/sec. The speed with which this process takes place is determined by the temperature in the upper strata, which on Venus has favored a high loss of hydrogen from the planet. The process might be two to three times faster than on the earth. Under these circumstances Venus' atmosphere, compared with the earth's, should have been gradually deprived of water, and this situation no doubt influenced the reversible reaction between silicates and carbonates, which in the presence of water at moderate temperatures goes predominantly in the direction of carbon dioxide binding and liberation of silicic acid; in the absence of water at higher temperatures the tendency is the opposite:

$$R{-}SiO_3 + CO_2 \rightleftarrows R{-}CO_3 + SiO_2 \,.$$

The more water that has disappeared from the surface of the planet, the smaller are the possibilities of maintaining a low temperature at ground level by the evaporation of seas and by removing the heat through radiation into the upper strata. The temperature as a result must rise, leading to an increase in photolysis, a loss of hydrogen, and a higher ground temperature and eventually to the state of desolateness which we can now observe on the plains of Venus.

How quickly can such a dehydration process take place? There is much to be said for the assumption that both Venus and the earth during the first and second billion years after the formation of the planet systems had approximately the same milieu; Venus was probably about 20 to 30 degrees warmer than earth and had about half its supply of water. Both milieus slowly developed organic compounds of varying degrees of complexity and increasingly intricate chemical structures from available carbon material on the basis of hydration and dehydration processes linked to sulfur and hydrogen sulfide of volcanic origin. Both developed gradually the first primitive forms of life and later more or less advanced anaerobic organisms. An evolution of this kind might have resulted—after another two or three billion years—in a situation in which additional supplies of water formed by volcanic activity would no longer be able to compete with the photolysis of water in the upper strata of the atmosphere. About one billion years ago a change took place in our planet through the evolution of photosynthesis which led to the rapid enrichment of the atmosphere with oxygen. As a result there was a pronounced photochemical formation of ozone O_3 from excited O_2. This ozone layer developed into a protective blanket against ultraviolet in the lower layers of the atmosphere and at the same time into an inhibitor of the photolysis of water.

The biologic enrichment of the atmosphere by an additional large supply of oxygen enables it to trap very effectively any photolytically liberated hydrogen and return it for renewed circulation as water. If this process had not taken place on earth, it could be questioned whether the milieu of life on our earth would not have become gradually like that of Venus with a continuous impoverishment of water through photolysis and hydrogen loss at an ever increasing rate. Any organisms would find it increasingly difficult to survive in such an environment, and they would eventually become extinct in a world without water and with an average temperature of 200° C and an atmosphere of carbon dioxide with a pressure of 5 to 10 atm.

This argument is based on simple and relatively reliable data. If it is true, Venus, for a period of about one to three billion years, would have had the same opportunities as the earth to develop life in an oxygen-free anaerobic milieu. The double amount of ultraviolet irradiation would have given the planet greater chances than on the earth to combine organic compounds to form living units, as long as there was still plenty of water. In the end, the severe drought must have set in and dried up all the labile carbon material which started to decompose and combine with the carbon dioxide liberated from the bedrock. Desolation was the final result. If these ideas are correct, we should be able to find fossils of complicated carbon material on Venus, such as coal, though probably not in very deep layers, which would be evidence of organism life that never got the chance for higher development.

How far did this extinction go? What evidence do we have to support the theory that no life whatsoever exists on Venus at the present time? From our knowledge about organisms on the earth, it seems unlikely that anaerobic organisms either would have had or have the same chances of developing as rapidly as aerobic ones, which have a much more efficient energy metabolism and a capacity to make advanced syntheses. All our knowledge of anaerobic life indicates that advances in one direction or the other are slow and that in a totally anaerobic environment chances of an evolution of conscious life are slight compared with those of aerobic systems. Given a long enough time and a fairly stable environment, conscious forms of life might develop in an anaerobic milieu, but on Venus the time was probably too short. But there are other possibilities which we may not have considered, maybe because they seem too unconventional from a terrestrian point of view.

The reports on the ground temperatures on Venus suggest that they far exceed what we consider the maximum for the existence of biochemical compounds as we know them, as well as of proteins and more complex molecular organizations; as a result, life on the ground level can be discounted without any reserve. We are used to seeing, observing, and imagining organisms as units growing and living either in the sea or on the ground, and we find it difficult to imagine other possibilities. But the fact remains that a small but significant fraction of organism life could function in the atmosphere as such. Our own planet gives an example of this.

If we disregard temporary visitors to our own atmosphere such as birds, insects, and the like, there are, nevertheless, some types of airborne cells, bacteria, lower fungi, etc., the metabolism of which is not destroyed but functions to a limited degree above ground level. Rainwater analyses reveal that organic substances are dissolved in raindrops and that a multiplicity of seeds, spores, and cells of all kinds are carried from place to place. It is not completely out of the question to imagine that such a system could be present on a large scale in another planet where the milieu of life is a dense, turbulent atmosphere, full of moisture, with a temperature around 0 to 30° C, with a certain concentration of substrates of a versatile carbon compound nature, and with two conflicting systems in operation, one in which the destruction of complex structures to simple compounds takes place and one in which a recombination takes place: resynthesis and reorganization. I have talked a little about this in earlier chapters, but I should now like to develop this theme further.

The more one thinks about this problem, the more one realizes the possibilities of certain forms of life even on planets with a markedly high ground temperature: life—organisms—bound to the atmosphere. For those organisms on our earth that have as their milieu the bottom layers of lakes and seas, the surrounding water is, relatively speaking, an atmosphere, an overlying region, the inhabitants of which can be regarded as atmosphere-bound. The situation is the same for us and other ground organisms; the atmosphere above us is an overlying region with possibilities of life in an atmospheric environment, although *our* relatively thin gas envelope has not especially favored the evolution of specialized airborne organizations of life at the cellular level.

The conditions on Venus, however, are highly favorable for the development of anaerobic, atmosphere-bound life. Let us relate this discussion to the conditions that exist at 30 km above ground level. There is carbon dioxide of 0.1 to 1 atm pressure, a water content which corresponds to 5 per cent or more of the amount on the earth, nitrogen in safe amounts (that is, a few per cent), and minerals whirled up from the ground surface which slowly give off soluble components under the influence of the raindrops. The temperature at 30 km high would probably be around 0 to +30° C. Moreover intense sunlight—twice the intensity of that on earth—which splits water into hydrogen and oxygen. *All this provides a situation reminiscent of photosynthesis in chlorophyll-containing plants on the earth.* This could take place in another milieu, on Venus, probably under much more Spartan conditions.

This suggests that some of the cloud particles which we can observe in the upper strata of Venus' atmosphere *may* be particles with catalytic activity—or organisms at the particulate level, if you like—with a capacity of reproduction, a short life—in brief, something living.*

The two regions between which the game of destruction and regeneration can be played are clear. The destructive region is the hotter layers of the atmosphere toward ground level, where organic compounds of all kinds are chemically simplified and broken down into fragments which will tend to recombine in a cooler milieu. The regenerating, constructive counterpart is the area at about 30 km above ground level, where the temperature favors the recombination of reactive molecules and where photochemical reactions of all kinds can occur in the dense atmosphere with its carbon

* Since this was written (January 15, 1961) the same conclusions about the Venus clouds have been suggested in an article in *Science,* April, 1961.

dioxide content which facilitates the organic-chemical complication of the molecules, helped by the hydrogen which has been split off from the water. It is quite possible that in the course of billions of years some of the participating units in this cycle of destruction and recombination have acquired forms which give them all the properties that we associate with organisms.

It seems fairly probable that evolution on Venus—which on the whole was similar to that on earth during the first billion years—allowed carbon compounds to form many complex molecules and the development of primitive cells of varying kinds. In the subsequent heating-up and dehydration of the milieu, only those cells or cell organizations which could continue to live in an airborne form in another milieu could survive. This would be in the cooler strata, which have a pronouncedly dense atmosphere, greater turbulence and activity, and an increasing possibility of sustaining organisms driven by the weather from the ground—a milieu whose low pressures and dense atmosphere represent the only possibility of a fairly undisturbed existence for the weather-beaten, reactive, water-starved life. In these centers of cyclonic activity primitive life of a microtype could have developed to a certain level—an evolution which would be characterized as a flight from the ground—but keen observers on the spot could interpret this evolution as being similar to their own situation—that of an evolutionary drive in a painful situation.

The inhabitants of the isolated ground base with its greenhouses and its slowly developing oxygen and food supplies speculate about these things every once in a while, as they look out at the restless and weather-driven dark skies during their brief periods of leisure. What are the chances of finding high in the stratosphere something similar to the extremely complicated fragments of carbon compound residues which are found scattered here and there in the banks of the sand dunes?

Anyone who tries in 1965 to summarize all the data on the planets is likely to push his conclusions to the limits of what is possible. It is not at all certain that our neighboring planet has been able to preserve any organism life from the predehydration period. Our analysis of the planet's surface conditions can perhaps give us only a clue about the ways in which organisms can exist and reproduce. The important thing is that the milieu does not necessarily have to be made up of seas or watercourses or water-irrigated areas, since an evolution similar to that which has occurred on earth could possibly take place in the higher strata of a dense atmosphere, if, of course, there were suitable temperatures and certain conditions favorable to destruction and recombination of carbon compounds.

As far as the permanent colonization of Venus is concerned, any measures that could produce a more moderate ground temperature should be utilized whenever possible. One suggestion worth considering would be to implant—many thousand years before such a colonization could start—resistant cells, related or similar to the Nostoc type. Such cell types, specifically adapted for life in an atmospheric milieu, if they had access to water, carbon dioxide, *and* nitrogen, and a certain amount of mineral salts, mainly phosphates, that had been whirled up from the ground, would have a chance of growing and reproducing in intense sunlight and would also contribute oxygen to the upper strata of the atmosphere. If a concentration of oxygen took place, it would form an ozone layer which would cut down the natural tendency of the photolytic hydrogen atoms to escape.

The result would be—one hundred thousand years later and after frequent injections of algae—the cooling of the ground's surface to a temperature below the 100° C mark, at which desert plants that are very economical of water could be planted. This would increase the oxygen production on the planet to the point at which respiratory organisms could be introduced in the cycle of life. Then, why could not man live there?

That man could colonize Venus is certainly and absolutely possible. The preparatory work might take a very long time—from about one hundred thousand to one million years—but efforts to achieve this may prove to be fruitful. In what ways? We might develop a new type of human being, specially trained to live and survive in a hot climate on a minimum of oxygen, a minimum of water, and an aggravating excess of carbon dioxide. They would be like a kind of Cytherea's Bedouins, making the best possible use of the rare oases which here and there would form an asylum for the struggling vegetation. It ought to be possible for our descendants to discover methods of extracting bound water and of increasing the ground vegetation. It will not be easy, and the sun-tanned descendants of the many generations who will have made tillable land of ground that just now is unknown and inaccessible will certainly develop legends about the pioneers who landed in dust storms, carbon dioxide pressure, and hideous heat.

CHAPTER VI

MARS: THE PLANET WITH POSSIBILITIES

On an astronomic level the differences between Mars and Venus do not at first sight resemble classical symbolism. Venus is hardly young, attractive, cool, and beautiful, except perhaps in its dazzling outer appearance, for at heart the goddess is frighteningly dusty, turbulent, full of dark violence, and unbearably hot. Mars on the other hand is rugged and open, and its cool dry surface and its undramatic season changes suggest an even temperament. In truth neither Venus nor Mars is quite young any more; the former is in an emotionally difficult transition period with a gentler future awaiting; the latter has had its share of strife and struggle and is now gladly entering the retiring age with a good deal of philosophical serenity. So far we have in a way made the astronomical data fit the classical descriptions of the two planets, and now without any further comments we can go on to analyze our other neighboring planet Mars, where most of what has happened happened a very long time ago.

The data that have been established with accuracy are set out in Table 10. First, we can calculate from the data that the gravity at ground level on Mars should be around 283 cm/sec^2, which is 0.38 of the value of 981 cm/sec^2 for the earth. The average distance from the sun, which is 1.52 times that of the earth, naturally means that Mars must receive considerably less sun radiation per cm^2 of ground surface, approximately half of the earth's radiation. The revolution time is very similar to that of the earth, being only some 41 minutes longer; so Mars has a day-night pattern that would be familiar to a visitor from the earth.

Data on the atmosphere and ground conditions have been obtained from observation, and these will be of great value to an expedition landing on Mars within, say, the next hundred or even fifty years. The thin atmosphere on Mars allows us to observe the

TABLE 10

Data	Absolute Values		Relative Values (Earth = 1)	
	Mars	Earth	Mars	Earth
Mass	0.613×10^{27} g	5.975×10^{27} g	0.11	1.00
Radius	3390 km	6371 km	0.53	1.00
Density	3.8 g/cm^3	5.5 g/cm^3	0.70	1.00
Orbit radius	228×10^6 km	149.1×10^6 km	1.52	1.00
Revolution period	687 days	365.256 days	1.88	1.00
Rotation period	24 hr, 37 min	23 hr, 56 min, 4.1 sec	1.03	1.00
Solar constant	0.92 $cal/cm^2/min$	1.92 $cal/cm^2/min$	0.5	1.00

ground surface so thoroughly—in contrast to Venus—that apart from the moon Mars is the best known astronomic object in our planetary system.

To begin with the atmosphere, data acquired by spectroscopic observations provide information about carbon dioxide and water. The carbon dioxide content has been estimated as twice that of the earth's atmosphere, that is, $0.00044 \times 2 \times 0.38 = 0.00033$ local atmospheres. The water content in the atmosphere is at the most one-tenth of the earth's, which would give a partial pressure of $0.1 \times 0.005 \times 0.38 = 0.00019$ atm. Oxygen is present in what seems to be one-thousandth of the amount in the earth's atmosphere. The local oxygen pressure would be approximately $0.001 \times 0.2 \times 0.38 = 0.00008$ atm. Taken together, the carbon dioxide, water vapor, and oxygen would produce a pressure of about 0.0006 atm, that is, approximately one two-thousandth of the earth's pressure.

We must take into account, however, the presence of nitrogen and argon in the same way as we argued for its existence on the earth and Venus. Nitrogen is certainly present in the form of molecular nitrogen, N_2, in quantities which dominate the atmosphere of the planet. It is difficult to quote a figure for the partial pressure mainly because of the difficulties of measuring at a distance something which does not show any absorption bands in its basic state. We must calculate by indirect reasoning and our conclusions imply that the low gravitation of Mars—with a minimum velocity of 4.2 for departing molecules as compared with 11.4 on the earth—must produce a much higher leakage rate for nitrogen and argon compared with the earth. Molecular nitrogen has a molecular weight of 28, and argon has a molecular weight of 40. Nitrogen, therefore, is more likely to leak out than argon. Carbon dioxide has a relatively high value on Mars, and carbon dioxide has a molecular weight of 44; so it seems probable that nitrogen with its molecular weight of 28 and argon with its even higher value of 40 would be able to remain, even if the local conditions favored a certain leakage of all these substances from the planet. An amount of nitrogen corresponding to one-tenth of the amount on the earth is probably a minimum value. This would give a partial pressure of $0.8 \times 0.38 \times 0.1 = 0.03$ atm.

Argon is easier to calculate, for its formation from potassium—$^{40}K \rightarrow {}^{40}A$—is independent of time and place and dependent only on the amount of ^{40}K deposited in the bedrock. We can assume—without any hesitation—that the distribution of potassium is equal to its distribution on the earth. As a result, the same amount of argon per cm^2 should exist on Mars, or possibly a little less because of the lower velocity of departure, which, however, at the present temperature influences molecules of a molecular weight of 40 only slightly. The result would be $0.01 \times 0.38 = 0.0038$ local atmospheres.

To sum up: The local density of the atmosphere should give an external pressure at ground level of between 0.03 and 0.1 atm, with nitrogen as the predominating element, followed by argon, which should amount to 1 or 2 per cent of the total pressure. Other elements contribute only insignificantly to the total pressure: carbon dioxide about 0.04 per cent, water vapor 0.02 per cent, oxygen a maximum of about 0.01 per cent. This estimate is not, of course, definite, but it should mirror the conditions on Mars to a reasonable degree. Data comparing the atmospheric conditions on Mars and on the earth are given in Table 11.

Determination of the density of the atmosphere on Mars will be of extreme significance for the first explorers. If the external pressure is 0.3 atm, this would correspond to the conditions at the

TABLE 11

Atmospheric Components	Mars Atm*	Earth Atm*
Nitrogen	Approx. 0.03–0.1	0.8
Oxygen	<0.0001	0.19
Argon	0.0038	0.01
Carbon dioxide	0.0004	0.0004
Water vapor	<0.0002	0.005
Total	Approx. 0.03–0.1	1.000

* Atm represents the local pressure; if moved to a planet of the size of the earth, the atmosphere of Mars would weigh about three times as much.

summit of Mount Everest, where a man equipped with an oxygen mask but with the rest of his body surrounded by the thin atmosphere can live and work, even though under trying conditions. If this were the case on Mars—with an external pressure of 0.3 atm —human beings should be able to remain in the open without pressure-proof clothes, provided they used conventional-type oxygen equipment. The low gravitation—only 0.38 per cent of the earth's—supposes a higher load of oxygen per capita and would at the same time provide a feeling of relative mobility, for an adult should weigh around 30 kg.

But a pressure of 0.3 atm is dangerously close to that of 0.1 atm, at which pressure the water in the body fluids starts to boil at a temperature of 37° C. If the nitrogen pressure in the atmosphere on Mars is down to the 0.1 atm level, bulkier clothes than the double-thick sweaters and well-insulated overalls of the Everest climbers would have to be worn for any length of stay in the open. A well-planned expedition would consider this problem in great detail and would be prepared for either alternative as far as protective clothing is concerned. In any case, the cold at night is so severe that some sort of advanced protective clothing is an absolute necessity for survival.

The temperature on Mars varies considerably, both as a result of the latitude and the day-night rhythm. The average temperature is 20° to 40° C below that of the earth's. Midsummer, a maximum temperature of +20° to +30° C for the day period has been recorded by long-distance measurements for the region around the equator. In the same area and at the same season the night temperatures are in the —60° C range, a somewhat severe temperature for summer. In the polar regions during the winter the day temperatures are probably —40° C to —60° C and the night temperatures disconcertingly low, probably —100° C. On the whole the temperature conditions would not be impossible to stand with suitable equipment; midsummer at the equator in the thin air would give a hardy Tibetan an illusion of being on home ground except that the scenery would have to be viewed through the visor of an oxygen mask.

The external conditions on Mars—pressure, temperature, light, and visibility—are therefore quite favorable for both landing maneuvers and for ground inspection by an expedition. The members of any such expedition would no doubt be well informed about all these conditions by means of data obtained from earlier robot space probes. An air of excitement would be felt as regards the uncertainty about the ground surface, both geographically and geologically, and the possibility of finding some endogenous form of life on this desolate planet, illuminated by a black and distant sun.

As the expedition approaches the planet, the surface contrasts which have been observed from the earth for a hundred years will become more pronounced. Red-yellow areas stretching over thousands of kilometers stand out in clear contrast to the gray-green zones, forming a strange pattern, which, toward the poles, is gradually transformed into well-defined regions of clear white, which look like water or hoar-frost or perhaps snow. The two satellites

Phobos and Deimos can be seen moving around the planet, their revolution times being 7 hr 39 min and 30 hr 18 min respectively. They are both quite small, with diameters of 15 and 7 km respectively. The former orbits at a distance of less than 9400 km from the mother planet, the latter farther out, at about 23,500 km, thus considerably closer to Mars than the Moon to the earth.

A faint veil around the planet, indicating a thin atmosphere, can be observed. What looks like a fog covers certain regions and slowly changes in character, suggesting wind and dust blowing over the cool surface, where little can be seen of rivers or lakes. The impression we get is of a cold desert with frosty mists and sand dust. But we will soon be able to see the planet in greater detail as the spaceship enters its landing orbit and puts on the brakes in the thin atmosphere, with its extremely good visibility compared with that on earth—not to mention the conditions on Venus. The landing place for man's first visit to the planet has been chosen in an area close to the equator in the summer hemisphere, near the border between a green zone and a brick-red zone. As the final stages of the landing operation are executed at a height of 20 km, the ground appears bare and without any prominent features. Continued braking makes it possible to land according to the methods which have been well practiced in the deserts on earth. Weighed down by its load of hundreds of tons of return fuel, the ship slowly descends toward the flat dunes and weather-driven *sastrugi,* tearing up a cloud of rough gravel in the pale morning light. Movement turns into rest and silence. The goal has been reached.

Outside the ship, which is still intensely hot from the friction caused by the descent, the temperature is a couple of degrees above 0° C. The external pressure level will decide whether space suits will be necessary in order to stay in the open or whether an inspection can be undertaken with only oxygen equipment and normal clothes. It is best to be cautious the first time and go out well bundled up until all the atmospheric elements have been analyzed. In this way there is less risk of unexpected calamities. Once you have got so far. . . .

The soft contours of the sand dunes fade away into the far distance in endless repetition. To the north, a darker contour stands out marking a break in the otherwise undisturbed stretch of red sand. A low hill near the dark shadow on the horizon seems to be a good place for a firsthand exploration of the silent region. The explorers wonder whether they will discover some life after a couple of kilometers or whether the gray-green substance will prove to be quite different from what they think it is. Preparations are made for the first long march in the pale sunlight of this strange world, where despite uncomfortable protective clothes and equipment one feels suspiciously light. Those who recollect the first walk on the rough ground of the moon are favorably impressed by the milieu on this desert planet. It could be worse. . . . Whatever we come across, this planet has many possibilities.

The possibility of building an observation base on this first visit depends on whether local water and oxygen can be concentrated by compression of the thin atmosphere. If these elements occur in a dilution corresponding to 1 to 1000 of the earth's concentration, then it will be relatively easy to make the most out of them by means of quite simple compressors and connected equipment for the separation of gases. Even by the most pessimistic calculations the planet ought to hold about 10^{11} tons of oxygen and as much water vapor in the atmosphere. This is not much, of course, in comparison with the 10^{15} tons of oxygen on the earth and more than 10^{18} tons of water in the atmosphere and the sea, but it should be enough for a while until one has a chance to consider what measures to take. The carbon dioxide content is relatively high; as a result local cultivation of plants under suitable protective covers with the utilization of sunlight seems a possibility. The amount of sun energy received is not so much smaller than in our subpolar regions on the earth (see the solar constants set out in Table 10 on p. 85). The desert sand, undrained by water to any great extent, will supply large quantities of salts to the first vegetation, which will be protected against the destructively cold nights by plastic coverings.

So far the situation seems favorable to the establishment of a base on Mars. In order to combat the extreme changes between the day and night temperatures, the main section of the base should be dug out a couple of meters below ground level. This should not be difficult, for the ground conditions are probably quite similar to those on the earth.

The winds will probably be quite mild movements of the air—most peaceful compared with the Antarctic storms on earth and the hurricane winds in the hot atmosphere of Venus. On the whole Mars's climate is highly suitable for man's expansion, for it is drier and less turbulent than Venus' and even, for that matter, the earth's.

If we want to set out our long-term aims, we must consider the possibility of creating—or re-creating—an atmosphere with such a high oxygen content that outdoor life would be possible without special equipment for individuals acclimatized to the place. At this point, however, we must make some qualifications in order not to make promises that cannot be fulfilled. The surface of Mars is approximately one fourth the size of the earth's, with a total area of 140×10^6 km^2. No water is present in the form of either lakes, pools, or defined rivers; there can only be a large accumulation of water in the polar regions, where the white ice caps occupy a maximum of 5 per cent of the total surface of the planet, that is, about 7×10^6 km^2. If these white regions should in fact be water—in this case, frozen—a 1 cm thick layer would correspond to 7×10^{16} cm^3, or 7×10^{13} kg. A layer with an average thickness of 1 dm would correspond to 7×10^{14} kg, a 1 m thick layer to 7×10^{15} kg. The latter would probably be the maximal average thickness, in which case the pressure per cm^2—if the layer were spread out over the entire surface (125×10^{16} cm^2) in either solid, liquid, or vapor form—would be $7 \times 10^{15}/140 \times 10^{16} = 0.005$ kg/cm^2. From a local point of view, this would imply 0.005 atm of water vapor, which could be converted into oxygen of corresponding pressure. But we need about 0.1 atm of oxygen to be able to manage reasonably well, that is, about one hundred times more than we could probably find on Mars.

A practical person would point out immediately that the greatest difficulty in creating a suitable atmosphere is the low gravitation of the planet, which extends unnecessarily the thinning out of the gas envelope toward space. An atmosphere with a maximum depth of 100 m would be quite sufficient for human beings on the surface of Mars. By a series of calculations we can see that a 100 m atmosphere with 0.2 atm of oxygen at the Mars level could be created from 0.002 atm of water, that is, exactly the amount we would obtain through the utilization of the water on Mars to transform it into oxygen via vegetation! A thin, transparent film over the base which would prevent the ground atmosphere of concentrated oxygen diffusing out into the kilometer-thick gas envelope and which at the same time would be light transmitting is something worth thinking about. In those enclosed regions oxygen could be provided by growing plants in sunlight on carbon dioxide and water according to well-established techniques.

To put this idea into practice, some ingenuity is needed. If the external pressure is about 0.1 atm, the same problem occurring on the moon will present itself—that of making the light-transmitting domes stable enough to withstand a pressure difference of about 0.9 atm between the outside and the inside world. The domes will in any case have to be small, with a rather large distance between the light intakes. One radical solution might be to choose an upland site and maintain the inside pressure at a level of 0.5 atm or even lower—a pressure which corresponds to that on the tablelands of Bolivia and Tibet—and to let the population adapt to this. The pressure difference between the inside and the outside could then be lowered to about 0.4 atm, which would prove to be of great advantage to the builders of future cities. Theoretically, we could go down to an inner pressure of 0.3 atm with a double amount of oxygen, thus creating a pressure difference of 0.2 atm, which is probably the limit from a physiological point of view. If the external pressure should prove in fact to be at the level of 0.2 to 0.3 atm, a large-scale colonization on Mars would be technically possible. With a pressure difference of 0.05 to 0.1 atm—and ideally none at all—an extremely simple light-transmitting screen could be set up between the oxygen-rich interior and the inert mixture of nitrogen and argon on the outside.

In this case double plastic layers could be set up over the base at a height of between 10 to 50 m, with a small excess pressure of carbon dioxide between the plastic layers to prevent any leakage of oxygen into the outside environment. External atmospheric strains in the form of wind are slight on Mars, and any temporary leakage could probably be repaired quite quickly because of the low pressure differences. So a local settlement could slowly grow up over the sandy plains, protected by plastic covers that would preserve the things important not only for the pioneers but also for their successors.

One problem will be the frost that will settle every night on the plastic covering, especially during the winter months; but by the time a large-scale settlement will be made, the special technology necessary for construction on Mars will have gone through its trial-and-error period, and more intricate problems than this will be solved.

What we have just been discussing must be regarded to some extent as a digression, or a practical vision if you like, in order to try and work out what opportunities there are for man on Mars, irrespective of whether it offers anything other than dryness, cold, and oxygen deficiency.

The question we must consider next is, of course, whether the planet has any endogenous life which could enrich the milieu in any way for the arriving strangers. This question is fundamentally important in itself—regardless of its practical consequences—and there are special reasons why it has not been brought up until now.

The classic question of life on Mars—not to mention any possible "advanced" inhabitants—can now be seen in proportion. This theme has been developed in fantastic ways in many kinds of literature, from H. G. Wells's *War of the Worlds,* to a section in Olaf Stapledon's *Last and First Man,* to Ray Bradbury's *Martian Chronicles.* But the more the problem of life on Mars has been studied seriously, the more difficult it has become to accept the existence of any form of conscious life in the conditions that now seem to prevail on the desert planet. What seems most probable in the present situation is something quite different from both H. G. Wells's and Bradbury's beings with their anthropomorphic features and their technical knowledge and aesthetic insight. Stapledon's concept of life on Mars as being organized clusters of virus particles of some kind is probably the nearest approach to what is scientifically acceptable—perhaps more adapted to Venus than Mars. We can, however, start to analyze the chances for any form of life, using as a basis our knowledge of the properties of the atmosphere and the main features of the ground surface.

First, any organisms which could exist on Mars—we are thinking of plants—must have the ability to survive brutal changes of temperature of about one hundred degrees every day. The carbon dioxide and water supplies are no worse than for the desert plants in the Sahara or in Arizona. There is sunlight, but it is only half as effective as on the earth, and the night temperatures fall to about —100° C. The closest parallel that we can find on earth is certain kinds of lichens in the polar regions which have developed an extraordinary ability to survive the most adverse conditions. We find them as far south as 86° S. where they bloom for a period on temporarily snow-free rocks—an insecure milieu —in sparse sunlight at temperatures at or just above 0° C. They develop in this environment cell divisions and other cellular activities within a couple of weeks. The main part of the year they spend in anabiosis, that is, in a biochemical sleep in darkness with temperatures of —50° C or lower. When they awaken at some time during the year, they have available unlimited amounts of local glacier water, which their possible colleagues on Mars, at least in the equatorial regions, can hardly enjoy.

If we had to synthesize a plant with a maximum of tolerance it would have to be one that would combine the ability of the desert cactus to economize water and the ability of the Antarctic lichens to withstand cold and to use every opportunity that milder weather affords to start assimilation and cell division. Such a combination, a kind of arctic cactus, with hardy fungal cells and an ability to adapt to extreme upland climates, would have a chance of living and surviving in suitable areas on the poor ground surface of Mars.

Several research workers have paid special attention to a particularly interesting feature in the life of lichens on the earth. The American physiologist H. Strughold, in his interesting study *The Green and Red Planet,* discusses lichens' ability to accumulate oxygen within their own organization for the consumption of the organization itself. A lichen is composed of an organized conglomerate of fungal cells in the form of fine threads into which chlorophyll-containing algal cells that have the capacity for photosynthesis are woven. In a way a lichen is a piece of local life in which inorganic carbon dioxide plus water are assimilated into organism material, carbohydrates, fat, and proteins by the chlorophyll structures of the algal cells under the influence of sunlight. As a result the algal component of a lichen can reproduce and divide, but at the same time, part of the material is consumed by the non-photosynthetic component of the organization: the fungal cells, which play the role of animalia in life on earth as a whole. The fungal cells—through the participation of oxygen liberated by the algae—decompose the complicated material which *is* the algal cells and their leakage products, and the organization seems to function as a whole according to the fundamental equations which are valid for life as a whole on our earth:

$$CO_2 + H_2O \xrightarrow{\text{Light energy}} (CH_2O) + O_2 \quad \text{growing algal cells}$$

$$(CH_2O) + O_2 \longrightarrow CO_2 + H_2O \quad \text{food- and oxygen-consuming fungal cells.}$$

This arrangement of algal cells and fungal cells seems to be relatively autonomous. It can function independently in a completely inorganic milieu which supplies the necessary mineral salts and nitrogen compounds, and it could live continuously, provided the milieu holds some carbon dioxide and water plus some activating sunlight.

Strughold emphasizes one fact that most plant physiologists are clear about: the presence of a high content of molecular oxygen in the interspace between algal and fungal cells in the lichens on earth. In his discussion of lichens in connection with the conditions on Mars, he suggests that lichens in an extremely oxygen-deficient milieu with access to a certain amount of carbon dioxide and water should be able to develop an extreme thriftiness with all the gaseous components by local exchange within the frame of the organization. One of the chapters in his book, entitled "The Hidden Store of Oxygen," clearly depicts the possibility that lichens may grow on Mars and that the main part of the free oxygen in the atmosphere *could be hidden within the intercellular structure of the ground vegetation.*

So far we have discussed *one* way in which some hardy form of life with the ability to make the best out of the somewhat sparse substrate could exist on Mars. The question which we will have to ask sooner or later is, of course, whether there really *is* life. It is not enough to conclude that it must exist just because certain organisms which flourish on the earth might be able to survive on Mars or because we are so willing to imagine that it exists that it does in fact do so.

This warning is justified from many points of view.

It is important to differentiate—at a distance like this—between the reasons for speculating that organisms might exist on Mars and the data which inspired these speculations. The concept of living organisms—and preferably of organisms of an advanced and conscious nature—is latent in us all, so that every change in the appearance of our neighboring planets, every report about the smallest traces of water and carbon dioxide, not to mention oxygen, gives our subconscious love of adventure a chance to express itself as an immediate correlation of the observed data with the possible existence of exciting forms of life in these distant places.

The observations of Mars in the 1890's by the Italian astronomer G. Schiaparelli, which were later continued by the American P. Lowell—both first-rate professional astronomers—showed that the surface of Mars has seasonal changes, many of them of a remarkably rapid nature. Every once in a while these extremely difficult observations, which were made with the eye as the only observing and the hand as the only registering instrument, led to false conclusions. For example, the spot pattern on the Mars surface, observed by eye and drawn by hand, was regularized to some extent, certain contours were accentuated, certain details were made to stand out more clearly against diffuse surroundings. This patient and completely disinterested work was the basis for the maps of the Mars surface. The clear-cut lines and well-demarcated fields gave an impression of a planetary surface geometry which was too well developed to be the result of chance and which must therefore be the result of purposeful activity. As a natural outcome of this, the hypothesis was developed of conscious inhabitants on the planet who by their heroic and assiduous work had created conditions suitable for cultivation on the desert planet through something similar to a canal system.

It should be noted that what Schiaparelli called *canali* (meaning lines in general) because of the tendency of the eye to see a dot pattern as a system of lines can be interpreted by a non-Italian to mean *canal,* that is, a canal that has been deliberately cut by some form of organism. It is quite possible that language rather than thought has gradually prompted a number of amateurs and even professional people to discuss the possibility of inhabitants on Mars in all seriousness. Lowell, who has his place in the history of astronomy because of his calculations about the existence of the planet Pluto, which was discovered long after his death, published in 1907 and 1909 books in which he described the advanced life on Mars, with irrigation systems and dams and a quite extensively developed technology.

Photographic techniques have gradually become more advanced from the beginning of the twentieth century until the present, and as a result, the geometric dot pattern on the surface of Mars, which had been earlier recorded by the pioneers of astronomy, began to look like a rather irregular system of darker and lighter regions, which, as spectrographic measurements have shown, coincide with the contrasting gray-green and red-yellow areas. The fact remains that the dot patterns vary with the seasons. Toward summer the gray-green areas increase in relation to the brick-red ones, which has given rise to speculations about their connection with some form of growing vegetation. The lichen idea seems to be a possibility.

Comparisons of the amount of light reflected from the green regions of Mars and the amount from arctic tundras with lichens as their dominating component have yielded certain results in common. Both Americans and Russians have tried with great care to find the absorption spectrum which is characteristic of earthly chlorophyll in the green areas on Mars; if this could be found, it would support the hypothesis of endogenous vegetation. The results are not absolutely convincing as yet—it would be extremely surprising from many points of view to find complete agreement between the dominating color of the vegetation on the earth and a component of the darker regions on the surface of Mars—but no observations which are completely contradictory have been reported.

In 1958, an observation on a much firmer basis was reported about certain absorption bands in infrared from the green parts of the Mars surface. Using the 5 m reflector of the Mount Palomar observatory, W. M. Sinton has registered three absorption bands in infrared—of the wave lengths 3.67, 3.56, and 3.43 μ—from the dark areas on Mars; these bands were absent in the registrations from the lighter regions. Organic molecules with hydrocarbon linkages of different kinds, in short, CH-structures, have a resonance frequency in this region with 3.5 μ as a mean value. Plants on earth, observed by the same method, in special cases show two bands at 3.51 and 3.41 μ. These observations seem to give direct support to the assumption that the surface of Mars contains at least the chemical combination CH in the form of hydrocarbon compounds of different types. There can hardly be hydrocarbon compounds in the strict sense; it is probable that other elements such as oxygen and nitrogen occur as components of these molecules. In this case we have two alternatives: the observed organic compounds either are relics from a previous dominance of hydrocarbons and similar compounds, that is, a fossil manifestation of the ways in which the elements could combine at an early stage, as on Jupiter and Saturn, or they have been formed later on as a stage in the development of life. So they could now represent something that either has been life or possibly still is.

The most interesting feature is, of course, the seasonal changes in the color of the gray-green regions. During each Mars summer the fields grow darker as if a wave of something dark started close to the polar regions and then spread slowly toward the equator, a phenomenon which in certain areas has coincided with a darkening of the red-yellow regions. Some scientists consider this phenomenon to be a basically physical phenomenon, caused by an increased humidity leading to the absorption of water by certain dust-dry minerals. This would cause them to change color, just as fine sand is considerably darker when it is moist than when it is dry. On the other hand, it is difficult to understand why only *certain* regions exhibit this phenomenon. This implies that the sand in these areas is of a completely different composition from that in others. This, of course, would be quite possible in a static milieu but hardly in a dynamic one where the atmosphere is constantly moving and causing sand storms, even if these seem to be quite mild. The Estonian astronomer E. H. Öpik, who has studied in great detail the changes on the surface of Mars, considers that the temporary mists which have been observed for a long time are sand smoke resulting from the movement of dust in the dry atmosphere. He points out that all the regions which are now dark and vary rhythmically from one year to another would have been covered by a uniform layer of yellow dust—the color of the moving material—if the darker material did not have, and still has, a capacity to regenerate its original appearance after being covered by the yellow dust. In other words, to quote Öpik: "The green regions should long ago have been buried under a cover of the same dry material that covers the rest of the planet, if it did not have some living component with ability of regeneration."

If we attempt to sum up the arguments for and against the existence of living organisms on Mars at the present, we arrive at the conclusions listed in Table 12.

At this stage the argument between the pros and cons tends to become more and more heated; so we will break it off here while it is still fairly amicable. The most interesting point in this discussion is that of the observations of the 3.5 μ bands. Why are the substances which give these absorptions not covered by the dust which *must* fall continuously over the whole surface of the planet, year after year, aeon after aeon? How is it that certain regions which undoubtedly have a very dry and dusty climate show surface reflections of CH-compounds while others do not, when the dust from the lower layers of the atmosphere should cover the entire surface of the planet.

TABLE 12

Data	For	Against
The seasonal changes of the green regions, with at the same time a decrease of the polar ice caps. The change begins in the polar regions and moves toward the equator.	An increase in the water content of the atmosphere and the ground surface would provide a stimulus for any vegetation. A thaw in the polar regions would provide water for vegetation in these and adjacent regions.	An increase in the water content of the atmosphere and the ground surface would lead to a moistening of dry sand = change. On the earth vegetation begins to bloom at the equator and then moves toward the polar areas.
It has been demonstrated that water and carbon dioxide are elements in the atmosphere	There is sufficient carbon dioxide for vegetation. There is also sufficient water, at least for certain types of plants.	There is sufficient carbon dioxide for a certain type of vegetation, but insufficient water.
Comparisons of the spectra from green plants on earth and those of the green regions on Mars reveal certain similarities.	Lichens can survive in conditions similar to those on Mars, both as far as the cold and the severe drought are concerned.	Not proven. Although it is perhaps possible there is no reason to assume that lichens grow on Mars.
The absorption bands at 3.5 μ in the green regions cannot be observed in the red-yellow areas. These bands in chemical compounds correspond to the combination C-H.	The absorption bands indicate the presence of organic molecules, compounds of a carbon-hydrogen nature, probably constituting parts of living organisms.	The absorption bands certainly indicate the possibility that CH-compounds are present, but these do not necessarily have to be part of something living at the present time.
Mars with its lower average temperature has had a chance of preserving organic compounds.	The milieu of Mars at an earlier stage was favorable to an evolution similar to that of the earth. The temperature was higher than it is at present.	Even if Mars was as hot as the earth, it probably "degassed" much more rapidly than the earth as a result of its lower gravitation.
There are traces of oxygen.	Oxygen is produced by photosynthetic organisms.	Oxygen is also produced by photolysis of water in the upper strata of the atmosphere.

TABLE 12—*Continued*

Data	For	Against
Organic compounds, carbon dioxide, water, and salts provide some of the essentials for the evolution of life in the form of organisms.	These essentials have been utilized on Mars as well as on the earth.	Nonsense. There is no evidence so far to support this theory. The earth has probably had special advantages, e.g., an excess of water.
During earlier periods water was present on Mars to a greater extent than at present.	The chance of advanced life was greater a long time ago than it is now.	Possibly, but has life in fact ever developed?

These are questions which may be answered during the next few decades by distance observations and space probes. The decisive stage will be that of man's confrontation with the ground surface and his actual inspection of the green material, which so far has revealed few of its secrets by long-distance observation. If it is something living, it will have to be extremely hardy, and future horticultural experts may be able to use it as a starting point in their attempts to develop an improved vegetation able to support a future population on the meager land of the tundras.

The sun slowly descends toward the horizon, and its rays accentuate the color of the desert land. The first explorers returning from their expedition to the dark, gray-green something, are somewhat confused as they brood upon what they have seen and have brought back for analysis. The first carbon compounds found in this quiet desolation do not indicate any especially pronounced metabolic activity. It is getting cold, and soon the night of inactivity and chemical sleep in arctic temperatures will set in. There is a crunch of sand under steel soles. Everybody longs for the relative comfort of the base. Although one or two think about what has happened, most of them are too overwhelmed by the experiences of the first day to do more than think about food and sleep. Tomorrow is another day. The sand is blown about by the evening breeze, the outlines of the spaceship stand out in the twilight. The planet is aroused to a new life.

For man life on Mars will take on certain special features which will shape daily life and work. First, he will have to get used to a low ceiling, literally speaking. The artificial atmosphere under the double layers of plastic covering will be limited to a certain minimum height in order to economize on oxygen and water. He will regard a trip "outside" much as we now regard the newly discovered sport of diving with oxygen mask and fins, the use of which has made us familiar with the shallower regions of the seas—a world outside the traditional surface world. The life "inside" as against the life "outside" will soon be reflected in man's thought and actions and later in recollections, legends, and poetry.

"Inside" will be synonymous with security, communications, routine work, a strictly regulated way of life where rigid laws govern the use of water, air, and waste material. The economy necessary for survival for this implanted life will gradually affect the mentality of the Mars population and its manners and customs in general. Ground transport both on short journeys "inside" and longer journeys "outside" cannot be based on conventional, oxygen-consuming fuels. "Outside" oxygen consumption is a technical impossibility, and "inside" it should be kept to a minimum. An ordinary car from the earth would appear to be criminal on Mars. Electric vehicles with efficient storage batteries for short journeys—they will weigh only 0.38 of their equivalent on the earth—seem to be the best solution.

The planet is relatively small, with a circumference of 21,000 km and a surface area of only one-fourth of the earth's; so air-proof electrical trains would seem to be the best means of transport for long-distance travel around the planet's surface. Railway engineers on Mars will certainly not encounter any very great physical obstructions in the construction of long-distance railway lines. Only for distances over 10,000 km will it be economical to use jet-propelled airplanes. The fuel will be expensive to begin with, especially since it must contain oxidants of a kind similar to those of our intercontinental ballistic missiles.

Life "inside" will always mean being together with other people, whether one likes it or not. In the early stages of development at least, the luxury of having roofed-in recreational areas out in the country will be too expensive. Anyone wanting to withdraw from the community for a time will only be able to go for a few hours of quiet contemplation—until the oxygen gives out. Two young people who are romantically interested in each other will find it difficult to arrange a rendezvous. There are too many people "inside"; "outside" there are the difficulties of trying to express one's feelings alone in the desert dunes in oxygen masks and protective clothing. Oxygen will probably be rationed, so that a certain thriftiness will have to be observed during the week in order to make a Saturday night date possible. However, the inventiveness which at all times has been typical of such situations—*Amor vincit omnia,* after all—will take care of things even on Mars, and as time goes by it may be recorded in the romantic literature of the desert planet. Venus will influence Mars, as it were.

CHAPTER VII

OUTPOSTS

At the moment Mars and the moon seem to be the only planets on which we, with our present resources, will be able to create a permanent settlement. From the communications point of view the moon is preferable because of its proximity, even though we must rely on the fact that future colonists both on the moon and on Mars will have to manage on their own to a large extent and live off the country, so to speak—for, whatever technical advances are made, transport to these planets will remain an extremely costly business.

In the long run Mars provides conditions more suitable for a permanent colonization than our neighboring satellite. The existence of an atmosphere, even if it contains only minute amounts of water and carbon dioxide, a lower variation in day and night temperatures, and a certain amount of protection against excessive short-wave radiation—all these make Mars more attractive than the moon as a site for enterprising settlers. The moon has its advantages from some points of view for astronomers and as a base for future long-distance trips, but the cultivation of the moon will always be a somewhat artificial project. The climate *is* harder than on Mars, whereas certain areas on Mars from a long-term point of view could certainly be cultivated; that is, we would not only be able to develop technical projects there—what we take to mean civilization itself—but we would also be able to create cultural centers, which we have also been able to do on the earth every once in a while.

At present Venus is out of the question because of its high ground temperature and a turbulent atmosphere that does not even allow experienced astronomers to make long-distance observations. Chemists on a short visit there might find many things to interest them, but on the whole the external conditions of the planet must be changed radically if this hot neighboring world of ours is to become a place where people can live and not merely survive.

Mars could become a place of permanent settlement—Venus perhaps later—in about one hundred thousand years or so. The moon could become a permanent observation post on the way to the borders of the universe, but what about the other worlds within the planetary system?

The worlds located outside the area which has the Mars orbit as its outer boundary and the Venus orbit as its inner boundary can probably never be settled by human beings to any extent—not even if we think in terms of a vision. These outposts of the planetary system located in regions either of intense heat or of intense cold on the borders of emptiness might be manned, for different reasons, by observers who might become the lighthouse-keepers of our species on the coasts of the sea of the possible.

In order to decide which of these planets are at all possible as outposts, we can start by discussing which ones we could land on, not only at the present, but also in the future when we are technically more advanced. Table 13 presents two alternatives. The planets have been divided into these two main groups for several reasons. First, gravitational conditions and external pressure are of greater significance than temperature conditions as far as landing is concerned. On the giant planets the energy required to brake a spaceship sufficiently to achieve a relatively safe landing corresponds to several times the energy needed to launch it from the earth! Moreover, the same amount of energy will be needed to launch it for its return trip. Second, the giant planets are surrounded by a gigantic atmospheric envelope which has hydrogen, helium, methane, and ammonia as its dominating elements in the outer regions. The ground pressure on Jupiter has been calculated

TABLE 13

	Mass (Earth = 1)	Radius (E = 1)	Density (E = 5.5 G/Cm³)	Gravity (E = 1)	Mean Temperature	Mean Distance from the Sun (E = 1)
	Bodies with Well-defined Ground Surface and Possibilities for Landing					
A. Mercury	0.045	0.39	4.18	0.35	+320° ± 100°	0.32
B. J. III	0.026	0.39	2.42	0.25	−120° ± 20°	5.2
J. IV	0.016	0.36	1.87	0.16	−120° ± 20°	5.2
Titan	0.024	0.37	2.48	0.32	−150° ± 20°	9.5
Triton	0.022	0.35?	2.2?	0.3?	−220° ± 50°	30.0
Pluto	0.93	1.0?	5.0?	0.9?	−220° ± 50°	39.9
	Bodies with Ill-defined Ground Surface and Few or No Possibilities for Landing					
C. Jupiter	316.9	11.2	1.1	2.5	−120° ± 20°	5.2
Saturn	94.8	9.5	0.7	1.1	−150° ± 20°	9.5
Uranus	14.5	4.0	1.3	1.5	−180° ± 30°	19.2
Neptune	17.2	3.9	1.6	1.6	−220° ± 50°	30.0
D. Other satellites and planetoids can be included in this group						

as being in the region of several hundred atmospheres; this is probably true for the other planets. Third, the gas envelope in the lower strata of the atmosphere probably changes into a liquid, forming a deep sea of liquid methane and ammonia, with a sludge of ice and solid carbon dioxide and complete darkness. Nearer the surface of the planet this becomes more gluey, probably even solid, long before the bedrock layer proper is reached, which is almost certainly covered by a solid crust of atmospheric elements in a crystalline form.

To plan to land in such an inhospitable milieu seems, to put it mildly, impossible to think about even as a pious hope for the future. If any future spaceship should for any reason go off course and get pulled slowly but steadily into the gravitation of one of the giant planets, the situation for the crew would be a nightmare of cosmic dimensions.

The smaller planets, with a gravitational pull lower than that of the earth, seem to be much more attractive landing places. At least there would be a chance to get off the planet; there would be fairly tolerable ground conditions—although one's expectations in this respect are not too high—and there would be possibilities for the establishment of a permanent observation base.

Of the six planets which we have decided are possible landing places, Mercury represents one extreme of temperature, with a ground temperature on the illuminated surface of about +400° to 450° C at the most and with violent differences of temperature between the day and the night sides. The latter are cold, with polar climates within a temperature range of —120° C and —220° C. It is difficult to decide which is the more attractive.

Mercury, to begin with, has an area and a density which are slightly larger than those of the moon, and the amount of radiation received from the sun is approximately four times higher than that of the earth-moon region. As with the moon an atmosphere in the proper sense of the word is lacking, and the value of the reflection power for sunlight is the same as for the moon, that is, 0.07, which suggests a similar surface structure. Thus the ground surface should be made up of material similar to that of the moon, with rocks and dust-covered surfaces. The dust probably has the same low heat-conducting capacity as the corresponding moon material, and this may prove to be useful in the construction of a base.

Of vital significance for such a project is the problem of whether there is any relatively temperate zone on Mercury which is protected from direct radiation. The data available so far indicate that its rotation is identical with its period of revolution, which means that the same side faces the sun at all times—as the moon faces the earth. As a result there are *two* extreme milieus on the planet: one extremely hot and one extremely cold. An area in between these two extremes should be suitable for a landing place and later on perhaps for an observation base. We must take into account the fact that the planet in its orbit around the sun moves like a pendulum on its own axis—a motion similar to that by which not only one-half but nearly three-fifths of the moon's surface is exposed to the earth by a slow rhythmic movement around a set position. The choice of a site for the base will be a matter for personal preference in these circumstances, but probably an area where the sun only periodically throws the full force of its rays on the ground surface would be best. This will provide the observers with an equally long period in arctic darkness in which to recover from the fierce radiation.

The base should be dug out below ground level, and the roof of the base should be able to vary its reflective power between mirroring white and maximal black in order to maintain a tolerable average temperature in the central parts of the base amidst the extreme temperature changes. If it is possible to stay outside at all—well protected by an insulated space suit—it will probably only be during certain limited periods when the huge sun has disappeared behind the horizon and the cold of the night has not yet really set in.

The external topography is probably very similar to that of the moon. Gravitational conditions are approximately the same; ground material and transportation problems are completely analogous, with the exception of the additional difficulties of obtaining supplies from the earth because of the distance. It would be an extremely trying place on which to live.

The first task for any observers on Mercury would be to study the sun at close range, perhaps from different viewpoints than we can imagine today. To plan a permanent observation base on this planet with its great heat seems unrealistic even for the distant future, since the moon will probably offer the same opportunities to obtain fundamental data about the sun. Thus it would be unnecessary to travel 92 million km closer to the sun; however, we cannot know for sure whether there may not be some advantages to be gained by exploring Mercury. The chances of being able to land on Mercury are, from the point of view of time, much greater than those of being able to land on Venus, with its elements of uncertainty, even if from a long-term point of view Venus could more easily become the center for a more permanent colonization.

With the experience of life on the moon, which will gradually have become normalized, to draw on, the pioneers on Mercury will encounter a world of approximately the same type, but the external problems of temperature and radiation will be much greater. A sun four times larger than the one we see from the earth or the moon will radiate the dry ground, long deprived of any volatile components, with the intensity of a blast furnace. If any of the elements of a primary atmosphere should remain, it will probably be one of the noble gases, most probably xenon. Some volatile metals should be present in the atmospheric envelope; mercury is probably a safe bid, and also some of the heavier metal halides. There would not be very much, but it might be enough to provide an external pressure of perhaps a ten-thousandth of an atmosphere, which would give the local sunlight some sort of shade in all its overwhelming intensity. The pioneers who land in this godforsaken place will probably not care too much for the wild beauty of the landscape. In this desperately hot climate they will no doubt think longingly of the more temperate zones of the solar system, and even the Spartan moon bases will seem remarkably pleasant and the outposts in the cold desolateness of Saturn or Jupiter, or even Pluto, quite desirable. . . .

The question is, however, whether the crew of a spaceship that has just landed on J. IV, which orbits at a distance of 1.88 million km from Jupiter, will agree with the Mercury observers about the external comfort of this place, some 800 million km from the earth. The sun gleaming through the thin atmosphere is about the size of a pea, being only one-fifth its size as seen from the earth, and its rays have hardly any warmth at all. The external climate is like that of an antarctic winter on the earth or even worse; the instruments show a temperature level which remains stubbornly around —120° C. The ground surface is covered by ice with solid hydrocarbons, solid carbon dioxide, and conventional snow mixed up together. There are few contours, and the landscape resembles what we imagine the Nordic Nifelheim to be like—a cosmic hell in its most extreme form.

But there are some bright points. One of the aims of the expedition is to make the best possible use of the chemical potentialities of the place, which has been frozen for some five billion years. Observations from the earth show that the atmospheres around the Jupiter satellites are very similar to those of the mother planet, with a remarkably high content of hydrocarbons and ammonia in the outer atmospheric envelope. Both J. III and J. IV have local gravitational conditions which correspond to approximately half that of the earth, and at the prevailing low temperature the content of methane and ammonia should be substantial. Both planets probably have an external pressure at the surface of one or two atmospheres. In choosing a landing place in the outer neighborhood of Jupiter the satellite with a ground pressure that most closely corresponds to that of the earth will naturally be chosen. The composition of the atmosphere and of the frozen components of the ground surface could be utilized to create heat, and not only heat but a completely independent supply of power for a long time to come.

The combination of methane and other hydrocarbons plus carbon dioxide, water, and ammonia provides considerable opportunities for chemists. The organization of a reliable power source—it may be a compact atomic power plant brought from the earth or the utilization of hydrocarbons by oxidation at high temperatures—will open up a wide field of chemical industry. The supply of hydrocarbons is practically unlimited, and through the chemical industry we are already well experienced in the technical utilization of such material. Everything, however, will depend on a reasonably easy mobilization of any oxidants which may exist, probably mainly in the form of iron oxides; but at the present stage, we cannot know for sure whether these exist. A priori their existence seems probable, for iron and oxygen are universal components, which on our earth, for example, make up 5 and 46 per cent respectively of the exterior rock strata; in addition, iron is the dominating element of the inner core of the earth.

If there were easily accessible deposits of iron ore, oxidation of methane on a large scale might be possible. The conventional processes used on earth would pose some difficult problems for a newly arrived expedition, for the ore would have to be mined, and furnaces based on atomic energy, in which the methane and iron oxides react, would have to be built. This process—at least in the earlier stages of colonization—could perhaps be carried out by simpler methods. Compressed methane and other atmospheric components could, for example, be made to react by supplying heat with an ore deposit. Thus the oxidation would occur on the spot, and the carbon monoxide, carbon dioxide, and water formed by the reaction could be collected and stored for further use:

$$3CH_4 + 4Fe_2O_3 + \text{heat energy} \rightarrow 3CO_2 + 6H_2O + 8Fe\ .$$

In this way valuable iron could be extracted on the spot and also utilized for the production of suitable construction materials. Carbon dioxide and water would be valuable for use in the oxygen-producing greenhouses. Moreover, the basic materials needed for a modern organic-chemical industry could be thus obtained by thermic cleavage of methane and other hydrocarbons.

But how should the atmospheric hydrocarbons be utilized with the aid of the local oxidants in the form of metal oxides? Many ingenious variations of this theme have been thought up, all of which are aimed at obtaining the oxidation with a minimum of extra energy in the form of heat being used. The arctic climate will prevent any organized cultivation of organisms for oxygen and food from taking place above ground. In contrast to the other planets in the system, from Mercury to Mars, sunlight could not be utilized to any great extent in the cultivation of vegetation. The amount of light energy received per unit of area is only 4 per cent of that received on the earth. It is doubtful whether it would be worthwhile to expose the greenhouses to a minimal amount of sun radiation, especially in view of the additional difficulties of insulating them effectively. It would probably be better to rely completely on artificial lighting in well-insulated mountain rooms. Water and carbon dioxide are available in adequate quantities, and in a suitable light green plants would be able to supply the base with oxygen just as they do in other places. By this time techniques will have been tried out in so many places in the planetary system that they will have been perfected both for extremely hot and for extremely cold areas.

It will probably be possible to develop some kind of igloo existence on the larger moons of Jupiter. A satellite with a maximum atmospheric pressure of 1 atm at ground level is preferable, so that protective suits would not need to be designed to eliminate excess pressure inside the suit. Insulation against the external temperature is not an insuperable problem, even if it exceeds by some 50° to 60° C temperatures which we can stand easily in our own polar regions; however, the relatively low gravitation favors mobility and activity at the ground level.

The first explorers are faced with some immediate tasks: the erection of a base from which the central planet can be observed at really close quarters. From J. III and J. IV, Jupiter appears to be an enormous object; its rapid rotation, which takes about 9 hr 50 min, makes it appear an ellipsoid. It will be possible to observe in detail the turbulent gas envelope in the different latitude regions and also to study the mysterious red spot—which is far larger than the earth and has puzzled scientists for about one hundred years. It should also be possible to get some ideas about the radio signal noise that is constantly emitted from Jupiter and to see whether it is connected with the atmosphere or with the ground. Microwave reflexes sent to Jupiter from one of its satellites would give us a picture of the ground structure and its temperature that is now blotted out by the atmospheric envelope. Such observations would help us greatly in our understanding of the origins of the planetary system.

All these things would apply to observations from other satellites farther out in the system. From Titan, one of the ten satellites of Saturn and approximately the same size as the earth, the ring system of the central planet could be observed with a clarity that is so far undreamed of. But it is possible that Titan might be completely unsuitable as an observation base, since the high content of methane and ammonia in its atmosphere might make visibility poor. If this were so, S. VII, Japetus, with its radius of approximately 1000 km and location at a favorable distance of 3.56 million km from the mother planet, would be more suitable. The temperature remains at about —150° C, but this should not be an obstacle if bases had been constructed on J. III or J. IV.

Farther out in space it becomes colder. In the Neptune-Pluto region, the temperature is —220° C, which means that most gases are in the liquid or solid phase. The boiling point for methane is —160° C; at —184° C it becomes solid. Hydrogen and helium are the dominating elements in the atmosphere, although the characteristic absorption spectrum of methane is present in the huge gas envelope of Neptune. Insofar as it is desirable to make observations in this extreme cold, Neptune's big satellite Triton, with a mass 0.35 per cent as large as the earth's, seems to be well suited as a base, if Pluto is ruled out because of its distance. Pluto has about the same dimensions and mass as the earth, and its low reflective power of sunlight suggests barrenness or at least a very thin atmosphere. Those explorers who in the very far distant future will land upon this inhospitable planet, which is some 60 billion km from the earth, will see the sun as a bright star, with the scope of the disk hardly visible to the naked eye. The radiation energy received from the sun is approximately 0.07 per cent of that received on the earth.

The messages that will be exchanged in the future between the earth and its outposts will require a great deal of patience. A radio signal from Mercury will take two to ten minutes to reach earth and as long for the answer. The exact time will depend on the constellation of the planets at the time of the communication. A message to a base in the Jupiter region will take thirty-five to fifty-four minutes to reach its destination, and an answer to even a simple question will take from one-half to one hour to come through even if the observers on J. IV answer immediately. It will take about ten to twelve hours for an exchange of messages to be made with Pluto—a long interval in cases where an immediate answer is required. Nothing can be done about this, however, since the maximum velocity of the radio waves is the same, and always will be, as the velocity of light, that is, 300,000 km/sec. This satisfies our requirements on the earth, but for interplanetary long-distance calls it is irritatingly slow.

Patience, hardiness, self-reliance, and inventiveness will be essential characteristics for the pioneers in the outposts of our planetary system if they are to survive and, moreover, work. In every situation, they will have to stand on their own feet without being able to rely on outside help. For those who survive, what will the compensations be? They will not be able to obliterate their experiences in a quiet life on earth after their return. Anyone who has traveled to the outlying lands of our planetary system and visited these fiercely hot or desolately cold regions has experienced things which will leave their marks on future generations. How?

CHAPTER VIII

THE RETURN HOME

I have outlined, only very roughly, the chances of life in our planetary system, both for endogenous life and for the earthly life which would be implanted. It is clear that many of the problems I have discussed have no immediate relevance; on the other hand, it is clear that many data of current interest have not been treated. This whole presentation should be thought of as a preliminary sketch in which I give the impressions that an explorer who has just returned would probably give.

The main theme has been to determine which chemical factors have been operating, both on our earth as well as on Venus and Mars, as components of a dynamic system and whether they open up possibilities of life. We are now beginning to obtain an insight into the events which led to the formation of life as organisms on earth—about one billion years before our time. The fundamental theory is that the carbon material which accumulated on earth at the beginning of time was of a hydrocarbon nature and that its successive oxidation in an anaerobic milieu gave rise to polymers of more and more complex structure which finally possessed a catalytic ability to influence the environment in a self-organizing direction. In a milieu where extreme destruction and organized recombination can take place, this material has gradually reached such a degree of self-organization that its function as a system of catalytic units with an ability to utilize the simpler units of the environment for its own growth has been secured.

The starting point for the earth and Venus and Mars should have been the same from a chemical point of view. The later developments of chemical potentialities have been governed by the respective distances of the planets from the sun and by their size in relation to each other. Table 14 serves as a guide. In broad outline this table shows the factors which have operated to give our earth a unique position in the distribution of carbon and water. On both Venus and Mars a considerable amount of water has escaped, much more than from the earth, the results, however, of quite different factors.

On Venus the photolysis of water in the upper strata of the atmosphere has been responsible for the main loss of water. The rate of photolysis for the first billion years was twice that on earth. On Mars the low gravitation has favored the loss of water, but photolysis per se has played only a minor role. The earth, located between Venus and Mars, with its solar constant of about 2 and with its relatively large mass, has been well placed in order to conserve water and carbon in the form of carbonate-bound carbon dioxide as a result of its water content:

$$CO_2 + RSiO_3 \xrightarrow{H_2O} SiO_2 + RCO_3 \, .$$

To sum up: Venus has been able to retain gases, but it is so close to the sun that photolysis of water has taken place on a large scale and hydrogen has been lost. Mars has been far enough away from the sun so that photolysis has occurred at a rate of only 0.25 of the corresponding rate on Venus; but its low mass, 0.11 per cent of that of the earth, has favored the loss of water as such and of carbon material in the form of lighter molecules. The earth has a favorable position a suitable distance from the sun, so that the photolysis of water is only 0.5 of that on Venus and water and carbon compounds have been retained, the latter mainly in the form of carbon dioxide. The oxygen formed by the photolysis of water should have been successively bound to the oxidizable materials of the ground surface, mainly iron-containing minerals at a

TABLE 14

	Venus	Earth	Mars
	Starting Points		
Position in relation to the sun (E=1)	0.72	1.00	1.52
Solar constant, cal/cm²/min	4	2	1
Mass (E=1)	0.81	1.00	0.11
Primary allotment of			
1. Carbon compounds per unit area	CCCC	CCCC	CCCC
2. Water per unit area	HHHHHHH	HHHHHHH	HHHHHHH
	The Situation One Billion Years Later		
Escape of lighter gases	+++	++	+++++
Photolysis of water: H→	++++	++	+
Remaining carbon compounds	CCC	CCCC	CC
Remaining water	HHH	HHHHHH	HH
Carbonate-bound carbon formed through the influence of water	C	CCCC	CC
Atmospheric carbon dioxide	CC	(C)	(C)

low oxidation level. The red areas on Mars are considered to hold iron oxides of the type Fe_2O_3.

If this hypothesis is workable, Mars should have approximately the same conditions now as shortly after the first heating phase following its formation. It may perhaps have had two to three times more water and carbon dioxide but not any more. In other words it will have a relatively stable and quiet milieu for about one-half to five billion years to come. Venus, one billion years or so after its formation, should have had similar conditions to those on earth at the same time after which time the local effectiveness of the photolysis began to differentiate between the conditions on the two planets. The Venus climate became warmer and warmer, and the life which two billion years or so ago had the same chance as on the earth to develop into organisms had to express itself in other ways. On the earth photosynthesis began on a large scale about eight hundred million years ago, with intense oxygen production and ozone formation in the troposphere which prevented further losses of water. On Venus the upper strata of the atmosphere must gradually have become the most favorable region for life, because of the excessive water photolysis in the lower strata and the increasing heat of the ground surface which made any return of life to it impossible. On Mars a slower tempo was present as a result of the sparseness of the substrates. Photosynthesis may be involved in any form of life on Mars, but if so, it would only be fairly recently and the chance of finding any advanced forms of life, such as organisms with a mobility of their own and consciousness, is rather poor. The same is true for Venus but for different reasons.

Chemical co-operation between a number of units of the organism kind in the upper atmospheric strata of a planet *may*, of course, cause them to develop into something quite complex, but the question is whether organized multicellular structures with individual or group awareness can develop in such a turbulent milieu, even over a very long period of time. To us on the earth, with a long period of relatively peaceful evolution in seas and on firm ground behind us, it seems inconceivable and even impossible in theory that conscious, organized forms of life could develop in other circumstances. On the other hand, we are bound to a conventional way of thinking about life. We think in terms of certain *manifestations* of life, although we are slowly beginning to realize that there are common denominators in the events that take place behind the façade created by organisms in activity. The next several hundred years of exploration of our neighboring planets wil probably give us, among other things, a broader view of the concept of life and consciousness, whether or not we meet these phenomena or whether our search only provides us with a reason to think more deeply about life as a universal phenomenon. In the long run this may prove to be most valuable for our continued evolution, perhaps more valuable than the data and materials that we bring home from distant worlds on our first journeys of discovery, which have been undertaken with great efforts as a grand adventure.

What at the moment often seems to be a very vivid experience—with positive and negative features—often appears quite different after a while. It will be the same when two explorers who have explored space widely meet again in the relatively quiet milieu of the earth and when the intensity of the immediate impressions dies away and leaves room for quiet thoughts and serene mediations. Our present technical developments in space travel and our confrontations with unexpected phenomena in our planetary system are beginning to leave their mark on our existence at home, and they will eventually form a background for a new expansion, this time toward the limits of what is universally possible. This is true for speculations about life as a whole within the universe, with its countless suns and their mysterious organization, as well as for the attempts to understand the structure and material anchorage of thought and consciousness within any advanced forms of life, wherever they may be encountered. All these things will become important after the return from a long voyage, when some who have seen and learned a good deal dwell unconsciously on the possibilities of seeing and learning more. Near by or far away beyond the limits.

MAN
ON ANOTHER WORLD

PART II. TOWARD THE LIMITS

CHAPTER IX

SIGNALS FROM SPACE

In any world where conscious life has begun to develop as a result of the tendency inherent in carbon compounds to combine in complicated forms, there is always the possibility that different forms of active consciousness may exist. In general, this consciousness may exist in any organism that has begun to utilize material and power from various sources in an organized way so as to develop some sort of latent mental activity. For man, with his one million years of evolution from the primate level up to his present state and his dazed cognizance of the possibilities of matter, the past one thousand years—the latest thousandth, a cosmic second—has led to the development of a primitive technical knowledge. This initiated first of all an intensive exploration of our own surroundings, then of our local planetary system, and finally of outer space.

We cannot but regard ourselves as being relatively undeveloped as conscious beings, with an extremely insignificant period of independent thought behind us; so it is clear that the idea of other worlds inhabited by conscious beings is something that does not require any advanced thought but is in fact obvious for any organism that has even simple aids for the study of nature within the local home region. By the time a conscious organism has acquired any insight into science or technology, the thought of exploring other worlds, first of all by increasingly precise instruments and later by space travel, is a natural stage of evolution, and sooner or later this exploration has to be realized, probably at a very early stage. From this point of view the chances that we will encounter a civilization within our cosmos, which has reached a certain degree of consciousness but which has not yet begun to experiment with the possibilities of interplanetary communication, are extremely small.

Our search for traces of life on our neighboring planets—and later farther out in space—will therefore be characterized by an either-or. If we find life, it will appear to be either relatively primitive and unconscious or conscious with a considerably longer technical tradition behind it than that behind our own civilization, which has passed the first expansive phase of evolution a long time ago, a phase which has taken us on earth a little less than two hundred years of technical preparatory work and which has been realized in practice during the last ten years—a minimal fraction of time on the universal scale. We are probably not alone in our efforts to broaden our local knowledge. Many have probably had the same goal and reached further than we have and have tried to send signals through space in their attempts to establish contact with other worlds in order to exchange data on life and its problems at an advanced level.

In the universe that we are able to survey at the present time we can register more than ten billion star systems, galaxies, and star clusters, probably only about one-tenth of what actually exists beyond the scope of our primitive instruments. 10^{11} organizations of suns of varying size and power-generation, with each organization containing an average of 10^{11} sun units, makes a total of 10^{22} (that is, about ten thousand million million million) luminous masses of intensely hot matter that are distributed about an unknown space where the intergalactic distances are counted in millions of light-years and where the limits of the observable, counted in light-years, are over two billion light-years away.

Within our own galaxy, the Milky Way, with its hundreds of billions of stars, there is an average distance from star to star in the order of about 10 light-years. We are situated somewhat at the periphery, where distances are counted in tenths of light-years. At the center, in the star clusters, distances are of the magnitude of light-days, even light-minutes, which might facilitate communication. The galaxy's greatest diameter is about 100,000 light-years, and its shape is that of a flattened-out spiral structure. On a cautious estimate, about one-thousandth of the suns in this local accumulation of matter in a state of thermic excitation belong to the same type as our own central sun. This gives us about one hundred million similar units. We know that a few of these units contain a planetary system, that is, a system made up of orbiting, cooled cosmic gravel. Are there any other similar units? The earth.

With the development of our astrophysical knowledge, theories about the shapes of the planetary systems are beginning to be put forward, and the theory that planetary systems are a general phenomenon is being considered together with the larger problem of the origin of suns. It seems that the formation of a planetary system is a part of a larger process that leads to the formation of a sun with its great accumulation of matter and energy. Planetary systems around every sun should be a rule rather than an exception, according to this theory. On the other hand, there are other theories, just as plausible, that state that a planetary system is a relatively rare occurrence, found only once, perhaps, in a million solar systems as a result of certain special cosmic events. If we assume the latter figure to be true—as a basis for calculation only—we arrive at the figure of a hundred planetary systems which have a central sun of our own type. If we calculate on the basis that all suns have a planetary system, we get a figure of one hundred thousand. Our own Milky Way, therefore, is only one out of ten—or a hundred—billion existing star systems.

The chance of finding planets within a planetary system that have the appropriate external conditions for life is naturally small, but not as small as we believe. In our own system there is a definite chance that life can exist within the region where the radiation is within the range of 1 to 4 cal/cm^2/min and where the planetary mass is large enough to retain volatile components within its atmospheric envelope. Thus any planet which is in the range of 0.1 to 10 times the mass of the earth and which has both water and simple carbon compounds can support life. If we make the pessimistic assumption that these criteria are only fulfilled in one-thousandth of all the existing planetary systems in our galaxy, we are nevertheless left with at least a hundred. If we accept the theory that a planetary system is a general phenomenon, we would have about one million. These would have the chance to develop life—some would have definite chances of developing conscious life—and with these we would be able to ask the questions we have formulated earlier in this book. For some planets this is all quite elementary, and for millions of years they have been aware of the possibilities of establishing intergalactic communications. If they make any attempts in this direction, it suggests that they judge the situation as clearly favorable for taking chances on a large scale.

Any explorations outside the home planet that any civilization in the universe might conceivably plan can be divided, as far as methods are concerned, into three types:

1. *Direct inspection.* Inspection of the planet chosen first by means of automatically controlled robots, which register data and telemeter the results to the home planet. After these probes come personal observations by means of a journey there and back by spaceship. This method would cover interplanetary research in which orders could be obtained from the home planet within hours, or possibly days. It also covers observations by spectral-analytic techniques during the early stages of exploration, before space travel itself can be organized. We are just beginning this project.

2. *Reception and transmission of identity signals.* On the whole this covers intergalactic studies. We may expect to obtain signals from objects up to a maximal distance of 10, or possibly even 100, light-years. The signals would be within wave-length range of electromagnetic waves, either light or radio waves of the short-wave type. Maybe also signals by laser techniques.

3. *Exchange of information material.* This covers work which will only be done in the far distant future, long after Project 2 has been started. The intervals between the transmission of a message and the receipt of an answer may be between 1000 to 100,000 years, at the best a few hundred.

Project 1 covers research at the home level and has already been outlined in previous chapters, but Project 2 has some interesting features which are worthwhile discussing, especially since this subject is currently becoming important to us.

This project is basically an attempt at intergalactic contact with objects located at a distance of many light-years. Our closest neighboring sun is α-Centauri at a distance of 4.3 light-years. There is another at a distance of 10.7 light-years, and two more at 15.9 light-years. The possibility that any of these suns has a planetary system which supports intelligent life is, of course, quite small. It might be profitable, however, to think about the technical arrangements that would in theory have to be made to transmit a message over this distance of between 5 and 15 light-years—an extremely short distance in terms of cosmic measurement.

To pick up such a signal at all on a planet of the size of the earth—in a form that is intelligible and which can be differentiated from regular cosmic noise—the signal would have to have a power of 10 w. We would need considerably stronger apparatus than we have at present in order to transmit such an amount over a distance of about 10 light-years in the form of intelligently pulsed signals. The parabolic radiowave reflector at Jodrell Bank in England, with a surface of half a hectare, can intercept signals of 2×10^{-13} w, which corresponds to a total incoming effect of 10 w directed toward half of the surface of the earth; but our transmitting capacity for the distance of 5 to 15 light-years is limited. We must calculate with complete precision the focus of the outgoing rays so that the transmitted effect is not spread over such a large receiving area that the small fraction that falls on any planetary object is less than 10 w. If the Jodrell Bank reflector had a focusing capacity corresponding to one ten-thousandth of a curve second, it could in theory transmit a message from earth with an outgoing effect of 10 w to a planet of the same size as earth located 5 light-years away, but this is an illusion. Such precision could only be achieved by a purely optic-astronomic instrument, such as the Mount Palomar telescope, in the focus of radio reflectors with their precisely shaped surfaces.

If we calculated with a much larger diffraction angle, for example, one whole curve second—a precision impossible to obtain—the outgoing rays would cover a surface 10^8 times larger than that of the target planet at a distance of 4.5 light-years from us, which would mean 10^8 times less energy received per unit of surface. Let us look at Table 15. The values underlined state our present performance ability as to focusing precision and transmitting energy per impulse. It can be seen that we are off by a factor of 10^8 with regard to the energy required at the present poor focusing capacity of 1000 curve seconds. If we could get down to a diffraction angle of one curve second, we should still be about one hundred times under what can be accomplished at the present time in

TABLE 15

Diffraction Angle of Transmitter Ray Expressed in Curve Seconds (3600 per Degree)	Surface Covered in Km^2 at a Distance of 4.5 Light-Years	Transmitter Effect Necessary To Transmit a Signal of 10 W to an Object with a Surface of 300×10^6 Km^2 at a Distance of 4.5 Light-Years
1000	3×10^{22}	10^{15} w
1.0	3×10^{16}	10^9
0.1	3×10^{14}	10^7
(0.0001	3×10^8	>10 w)

terms of energy per impulse. It would seem better to work for greater precision rather than the mobilization of larger amounts of energy for transmission. Precision here means something like a surface accuracy of ± 0.01 mm for a parabolic reflector with a radius of 100 m. Such a reflector, with an accuracy of 0.1 curve seconds and an energy of 10^7 w per impulse—the most that can be realized at present—could transmit the necessary 10 w to a receptor surface of the same size as the earth located at a distance of 4.5 light-years. If we could achieve in the near future outgoing impulses of approximately 10^9 w, which is a tremendous amount, the accuracy of the reflector construction could be brought down to ± 0.1 mm, which is certainly not beyond the bounds of possibility.

It is clear from these calculations that we cannot yet mobilize the necessary resources to send messages through the intergalactic space to the nearest solar system, but we may be able to do so within a hundred years or perhaps earlier.

A project like this naturally rests on the assumption that a target signal will be picked up somewhere, interpreted, and answered somehow. It is a game with probabilities in which the chances of success are small but in which any returns would be immeasurably important. The question is, How many objects in our immediate vicinity—within a range of 5 to 500 light-years—are occupied at this moment with listening and with transmission, which, after all, are criteria of a technically advanced conscious intelligence?

If one day we should happen to intercept an unusual series of signals of more than a few seconds' duration which suggest that some kind of intelligence lies behind the transmission, we can be certain that these rays have been deliberately directed toward us. Undirected radiation from a transmitter has no chance of reaching an object at a distance of more than a couple of light-years, not even with a huge amount of outgoing energy. A *fixed* parabolic transmitter with an accurate focusing mechanism can be ruled out because the rotation of the planet in question would cause the search ray to sweep continuously through space without any long period of contact with any particular object. A transmitter planet with a revolution time of 24 hr at a distance of about 5 light-years from us and with a fixed parabolic transmitter would sweep its transmitter ray through our whole planetary system in a maximum of one second, which is too short a time to enable us to pick up its methodical pulse pattern. If we should ever intercept an intelligent series of signals, this would mean that our sun and other objects in the signal path had been chosen on the grounds of the high probability that planetary systems exist in this section and that the probability had been judged to be high enough to warrant the waste of expensive energy in this direction. On the other hand, if we work out a pattern by which we can listen systematically, we find that there are only a few objects against which it is worthwhile to direct a large reflector. We have to limit ourselves to objects within a distance of about 5 to 15 light-years because transmission over greater distances seems—at least for us—to have little chance of success, even with all the technical resources that can be mobilized by a superintelligence with an unlimited supply of energy. Our choice is therefore limited to the immediate intragalactic neighborhood, which is represented by a narrow selection of suns located at a distance of 4.0 to 16.0 light-years (see Table 16).

Spectral class designations are an astronomic convention by which the chemical and physical status of a sun is indicated by a G_0 symbol, and the series O, B, A, F, G, K, M, N, and S with the suffixes 1 to 9 indicate sun evolution in general according to a certain pattern. We should concentrate primarily on a planetary system that has a central sun that is neither *too* hot nor *too* cool

TABLE 16

Object	Distance in Light-Years	Spectral Class (Our Sun = G_0)	Radiation Intensity (Our Sun = 1)
α-Centauri.......	4.3	G_2	1.1
α-Canis Majoris..	8.6	A_1	26.0
α-Canis Minoris..	10.0	F_5	5.0
ϵ-Eridani........	10.7	K_0	0.33
γ^2-Eridani A.....	15.9	G_5	0.36
γ^2-Eridani B.....	15.9	B_9	0.048
γ^2-Eridani C.....	15.9	M_{4e}	0.0012
α-Aquilae........	16.0	A_7	9.0

for life to have a chance to develop. This limits our selection to three that seem to be the most likely: α-Centauri, ε-Eridani, and γ^2-Eridani A. The first and last belong to a spectral class and a type fairly close to our sun. A project was started in 1962 at the radio observatory at Green Bank in the United States with the purpose of listening to short-wave noise from these objects, and special attention was being given to ε-Eridani.

The technique used is to try to discover characteristic patterns in the static short-wave noise transmitted from innumerable regions of our own and other galaxies where thermically excited hydrogen emits irregular amplitude variations in the 21.1 cm band. The incoming signals from the object studies have to be compared at short intervals with the surrounding general noise in order to discover, with the aid of modern data computers, whether the incoming signals from the goal region have some antistatic content. In other words, do the signals contain anything like a message such as: Here am I; where are you?

Any presumptive sender organism is therefore working on the basis that any fairly intelligent receiver has made arrangements by which they can listen in. The problem now is in what form should a series of signals be sent so that it would be clear to a primitive being within reach of the directed search beam that the noise is not static.

We can assume that any signal sent out from a planet where a certain scientific-technical knowledge exists will be extremely uniform, wherever it comes from. It is likely that all program directors reason according to the same criteria, that is, that the signal should be different from other noises in the universe and should reveal its basis of thought, so that it can be seen as an expression of life. The prime number series 1, 2, 3, 5, 7, 11, 13, 17, 19 . . . would be a simple signal that would indicate that the sender has some idea of mathematics and that he knows that the prime number series could not arise through any unconscious transformation of matter. The sender will know that if we are reasonably gifted we will already be aware of this and understand that the prime number series must be expressed in a universally current mathematical language and not in the form of our local decadic counting, 1, 10, 100, 1000, 10,000, and so on. Our—and any other sender's—choice would certainly be the binary system: 2, 4, 8, 16, 32, 64, 128. . . . The latter signal has the advantage that additional numbers can be transmitted as a series of dots and dashes. The first outgoing signal made up of the prime number series from 1 to 19 would therefore look like this:

o	o—	oo	o—o	ooo	o—oo	oo—o	o———o	o——oo
1	2	3	5	7	11	13	17	19

This signal would alert any listener and he would wait for the continuation with interest. It comes:

o—	o—o—	o——o—	o——o——	oo—oo—	o—o—oo—
2	10	18	36	54	86

This new signal tells us that the sender is introducing us to atomic physics, that is, to the number series which indicates the maximal number of electrons in a finished shell in the series of elements. It gives as an example the electron numbers of the noble gas series and therefore also their nuclear charges, from $_{2}He$, $_{10}Ne$, $_{18}A$, $_{36}Kr$, $_{54}X$, to $_{86}Rn$, probably in order to alert us that a chemically-oriented signal will follow:

```
o o—o— — oo—o— o—oo— oo— — o—oo oo—o oo— — — ooo— —
o o—o— — oo—o— o—oo— oo— — o—oo oo—o
o o—o— — oo—o— o—oo—
o o—o— — oo—o—
o o—o— —
o
o
o
1. . .20. . . . . . .26. . . . . .22. . . . . .12. . . . .11. . .13. . . .24. . . . . . . . .28. . .
```

$_{1}H$, $_{20}Ca$, $_{26}Fe$, $_{22}Ti$, $_{12}Mg$, $_{11}Na$, $_{13}Al$, $_{24}Cr$, $_{28}Ni$

This is clear. The information about the nuclear charge numbers for hydrogen, calcium, iron, titanium, magnesium, sodium, aluminum, chromium, and nickel at once gives an astrophysicist a clue about a certain spectral type, especially if the series is repeated in varying ways which correspond to the relative amounts of the elements mentioned in powers of two, with hydrogen as the dominating element. Translated into our earthly system of classification of sun spectra, the signal suggests a spectrum of a type between F and K, probably a variant of the G series. This would enable us to search by means of telescopes for objects which would fit the description and which would probably be quite near. If there is a sun of G type within a reasonable distance in the region from which the transmitted signals are coming, then it is probable that this sun has a planetary system, from which conscious life has just signaled a description of its central sun as an invitation to further correspondence.

If the signals occur with a regular interval of a certain number of hours, this would suggest that the interval designates the time taken for the planet to rotate. By careful observations over a long period of time, we should be able to establish that the time of the daily broadcast—by our measurements—is displaced by a certain amount according to a long-term pattern. A displacement of 5 min in each rotation taking 28 hr over a long period of, for example, 390×24 earthly hours would provide us with the information that the unknown planet has a rotation time of 28 hr in terms of our measurements, that it rotates a total of $28 \times 60/5$ times per orbit around its local sun, and that it has 336 local days during the time that the earth has 390 days. This information tells us that the planet must be located relatively close to its central sun. The size of the latter is revealed to a certain extent by its spectrum, and as it seems to be similar to our own sun as far as size and intensity are concerned, the planet should be about 100×10^6 km away from its central sun, that is, in an orbit which would correspond to a planet between Venus and the earth in our system. Its rotation time is similar to that of the earth and Mars.

If planet A receives a brief signal from planet B that contains information suggesting intelligence and knowledge of basic data of physics and chemistry and that also includes a statement of the main spectral features of the sender's central sun, planet A should formulate a sensible answer, if A is capable of doing this. The answer should be drawn up according to the same scheme as the message received in order to show that "your signal of xxoxxox" has been received and noted down. An answer in the form of the prime number series beyond 19 to 23, 29, and 31 would be a good idea to show that the incoming signal was comprehended. Certain peculiarities in the spectrum of one's local sun could then be used, for example, the element $_{21}$Sc, scandium, o—o—o, as an identification signal. Then A can transmit toward B for a long period, although with the knowledge that answers to their messages cannot be expected for 9 to 30 years at the earliest.

If in the future planet B picks up the response from planet A, a variant of the earlier signal which has been sent for perhaps 1000 years should be formulated and sent out as soon as possible. The prime number series could be extended to . . . 37, 41, 43, and 47 . . . and the sender's little specialty, element 21, o—o—o, could be transmitted acknowledging the receipt of "your signal of. . . ." Planet A after careful listening for 31.8 years notices suddenly that the incoming signal from B has altered; it now has the addition of o—o—o and perhaps some other local information. From this time interval it can be calculated that the probable distance from A to B is $31.8 \times \frac{1}{2} = 15.9$ light-years; then A can at once send a new signal to B which will tell B 15.9 years later the distance in space that separates the two planets. The field is then open for further exchanges of thoughts for a long time to come. Data about external conditions on the two planets, about their atmospheres and its dominating elements, will be transmitted relatively rapidly—that is, over a period of several hundred years—and will outline in broad detail the chemical milieus of the two different forms of life. Symbols for summation, subtraction, and other simple mathematical operations should, of course, be exchanged as soon as possible, as well as symbols for more complicated chemical structures, all of which could be transmitted, with a little ingenuity, as a series of signals of the o— type. The exchange may seem slow, but there is plenty of time. Perhaps in many places in the universe, in its endless emptiness with its intense play of forces, an exchange of ideas may have begun a long time ago, with more than two participants, perhaps leading to a more and more developed consciousness on the part of the participants. In the long run the dot-dash system will, as already suggested, bring about a primitive television picture of life on a distant planet. This is a continuing, definite possibility, even if the information gained may not be too overwhelming.

Yet from the dot-dash patterns a picture is slowly built up about the life and milieu of a distant world which will remain beyond reach for the next 1000 or perhaps 10,000 years. Some kinds of information, difficult to transmit as a series of symbols only, could be telemetered in the form of pictures. We would gradually—perhaps relatively quickly—acquire an insight into the external conditions of the organisms in this remote world who are sending data about themselves and their planet. It is even possible that the transmitting organisms on planets A and B could exchange pictures of themselves as a courtesy.

At first simple data about temperature levels, atmospheric components, radiation from a life-activating sun, and organisms in activity, organization, and conscious planning will be exchanged. Much of the information transmitted from either side will be incomprehensible except to the sender, but the theme will remain the same: an attempt to find a denominator common to the two worlds of consciousness which have established contact with each other in order to increase their comprehension, at an advanced level, of life.

Later on material of an informative character, such as that outlined in project 3, could be exchanged. We cannot exclude, from a purely technical point of view, the direct transfer of material, both inorganic and organic, once contact has been established between two advanced types of organisms at a reasonable distance, let us say within 10 to 20 light-years. This would be the first attempt at planned interstellary travel. The execution of such a project would require advanced technical skills, but its achievement is feasible.

Such a project would have to be based on the launching of a not too complicated probe which had instruments to adjust its course toward a given goal. In the vicinity of the sender's planetary system, the course could be controlled by radar from the home base; this is not too difficult to arrange. After this, adjustments would have to be carried out by a photo eye persistently facing the sun of its destination—for a period of 10,000 years for every light-year of distance, the speed being the same as we can achieve at present, that is, about 30 km/sec. By that time, however, new propelling devices based on ionic or photon radiation will be in operation, by means of which the speed of space probes can be accelerated to a speed of 30,000 km/sec. It is probably unrealistic to imagine any higher speeds, but a speed of 30,000 km/sec is approximately onetenth of the speed of light. With an interstellar distance of 10 light-years, the journey, even at this speed, will take at least a few hundred years in view of the relatively slow rate of acceleration up to such a speed. Deceleration and braking would also have to be put into operation at an early stage so that the entry into the planetary system of the goal region can take place at such a low speed that local maneuvers can be carried out within a limited region about 1 billion km wide.

As the spaceship enters the planetary system of its place of destination, a short-wave signal should be turned on to indicate its approach to those who await its arrival. Local radar instruments in co-operation with the perceptive and maneuvering mechanisms of the ship itself should be able to control the spaceship. At this stage it should not be too difficult to direct the spaceship to its place of destination, provided that everything has gone according to plan during its multicentury flight through space.

What should the material which will be waited for so patiently be? Here is a subject rich for speculation. How could we sum up our own life and evolutional history within the framework of all other forms of life that have taken place in our changing inorganic milieu over billions of years? We would only have a few cubic meters of space, perhaps less. So we could not send more than a few specimens out of the immeasurable variety of life forms represented by small organisms. These could be sent in a frozen state and thawed out carefully on the foreign soil in order to bear witness to the common factors that bind together all that lives and has had an opportunity of developing into an awareness of life. A little soil, some plants, a few insects,—at first sight they look strange, but on closer examination they are probably only variants of a theme which has been known for a long time.

We should also add a few microphotographs which when enlarged would give some idea of the beauty of a foreign world as it is appreciated by someone who lives there. We should also send some things that we consider to be typical of our own milieu, which might convey more of our personality and of the future possibilities of contact than all the signals and exchange of materials could ever do. All this may have been going on for a long time, long before we even thought of listening and waiting.

CHAPTER X

ADVANCED CONSCIOUSNESS

We are gradually becoming more aware of the possibility that conscious life could have developed further in areas which in time and space are beyond the limits that we can survey at the present time. The term advanced life that we have used from time to time in earlier discussions suggests a hopeful prognosis for the future, an unbounded faith that mental development is synonymous with advancement. Let us analyze this further.

Until we do hear and comprehend signals which can give us information about life in other regions of our galaxy—not to mention other systems in space—our speculations about the shape of conscious organisms and organizations remain naïve extrapolations of the possibilities of earthly fauna and flora, including ourselves. It is, of course, easiest for us to think in terms of our own situation today and to imagine that the future—one or two million years away—will be patterned on our own philosophy, peculiarities, and technology carried to extremes. The question is, however, To what extent is Homo sapiens, with his extremely short evolution into a conscious being, suitable as a starting point for a general extrapolation of the possibilities for the further development of a conscious organism into a more advanced consciousness? The basic material for such an analysis—to the extent that it can be carried out at all—is undeniably meager, and we run the risk, moreover, of limiting our thoughts about consciousness as a general problem to what is only one of innumerable forms of a universal tendency.

In these circumstances, it may be better to take a broader view and first attempt an analysis of the factors which are common to the evolution and function of terrestrial life as a whole. This will at the same time give us an orientation by which to consider the evolution of life as a general phenomenon. This means that we shall have to dive deeply into the biochemical world, paying special attention to the organizing factors which co-ordinate a multitude of chemical processes into integrated systems with specialized functions. If we follow this line of thought, we may perhaps be able, gradually, to acquire a basis for an initial discussion about the concept of conscious life and its universal possibilities of evolution to an advanced level.

Earlier in this book I have set out the arguments which suggest the existence of common factors in the chemical components of living systems. The central point, as I have already pointed out, is the potentiality of carbon compounds to exhibit variability and reactivity. As far as we can reconstruct the evolution that took place on our earth during the first billion years, it seems that a number of extremely complicated molecular aggregates with catalytic power, forerunners of our present enzymes, were formed gradually from an enormous multitude of molecular types, with carbon, hydrogen, oxygen, and nitrogen as the dominating elements. Development from this point was essentially a question of increasing organization and concentration of multienzymatic systems according to a general, basic scheme. Roughly speaking, this development was that from *simple systems in a complicated milieu to complicated systems in a more and more simplified milieu.* Translate this into chemical language and we get the following: A development of a few types of extremely primitive enzyme complexes in an extremely variable milieu of carbon compounds of all types into a multitude of complex multienzymatic systems in a milieu which is increasingly deprived of complex compounds, the final stage of which is cells in a milieu of carbon dioxide, water, and simple nitrogen compounds. Figure 8 outlines this evolutionary tendency in pictorial symbols.

From this slow process, which must *in principle* have functioned

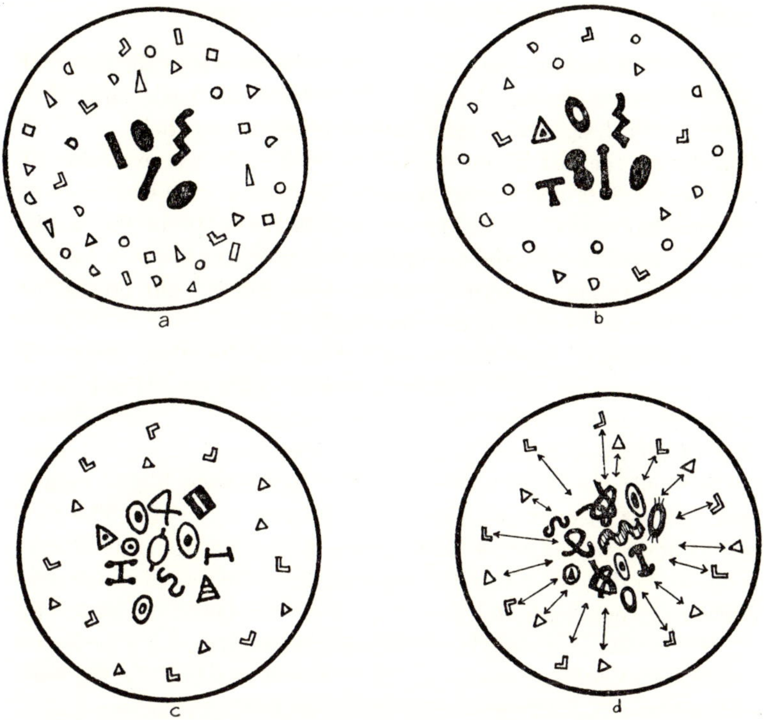

FIG. 8.—Symbolic representation of the evolution of life. The first stage, *a,* represents the state of things 3 to 4 billion years ago when the carbon compounds of the earth's surface were much simpler, but when there was a multitude of different types of compounds. At this time, too, high-molecular material was simple; there were only a few types, and they were not very reactive. In stage *b* the high-molecular material is more varied; the low-molecular material in the surrounding milieu has been reduced in quantity and number of types. This tendency continues for 2 billion years until the low-molecular material is almost entirely represented by carbon dioxide and carbonates; the high-molecular units are beginning to be organized into multifunctional units (stage *c*). Further evolution and complex developments lead to the stage shown in *d.* This symbolizes the results of the ability of certain units—via primitive photosynthesis—to utilize the carbon dioxide of the milieu for the formation of units of their own, for the growth of advanced units, for locomotion, that is, new complications in a milieu where living units are no longer bound to certain limited localities. This represents the primitive stage of the first flora and fauna in the pre-Cambrian—about one billion years before our time.

—and still functions—in parts of the universe where the external possibilities for the development of life exist, the basic units that we on earth designate as cells have developed, and their organization during the last billion years has resulted in the evolution of our present flora and fauna on earth. Discussion of the more complex mechanisms of evolution are outside the scope of this work, but it is important for the arguments that follow to know in outline the organizational pattern by which every cell is built up, as far as structure and function are concerned.

Irrespective of their variations in type and function, every cell is an organization of subcellular particles with specialized functions. The past twenty years of biochemical research have shown that of all the different kinds of particles found in oxygen-consuming cells the *mitochondria* and the *microsomes,* highly organized systems of enzymes, have quite different functions. The former contain the enzymes that remove hydrogen from the simple compounds that constitute the common denominators in the degeneration of complex material such as food substances of all kinds, as well as the components of the cells themselves. In a series of co-ordinated enzyme reactions, the hydrogen donated by the normalized material—for example, active acetate, pyruvic acid, glutamic acid—is transferred to other substances, or to oxygen, for the final oxidation of pretreated, simple material to water, with a continuous liberation of carbon dioxide. The energy gained is stored partly through connecting processes as phosphate compounds with an activating effect on slow-reacting material. This is a system that is based on the *decomposition of the main part of the undifferentiated material in order to gain energy for the synthesis of new, complex material and to make possible its organization into the supercomplex structure that is the cell.*

The function of the microsomes is to utilize the active phosphate compounds produced by the mitochondria for a certain, specific phase of this synthetic activity, that is, the synthesis of proteins, most of which have enzymatic power. Some of these enzymes will become parts of new mitochondria, and so we catch a glimpse of how the organization reorganizes itself.

The co-operation between mitochondria and microsomes is carried out under the influence of the material forming the genetic organization within the cell, often localized in an especially demarcated part of the cell nucleus. A discussion of this control mechanism could easily sidetrack us; it is sufficient to indicate that the influence of the cell nucleus is exercised by means of certain complex substances containing sequence-bound units, DNA, deoxyribonucleic acids, which somehow direct the protein synthesis and therefore also the synthesis of enzymes in the microsomes. The mitochondria provide the energy; the rest of the cell furnishes the raw material.

Figure 9 indicates the nature of the co-operation between the three different bodies within the frame of a cell. The activity within a cell is clearly much more complicated than this, but we can see that the co-operation of the different units, each of which specializes in some chemical function, gives the cell a possibility of existing as a complex organization in a less complex milieu at the same time that the organization is continuously regenerated through the transformation of milieu components—in this case, food substances. This is true for all the bodies that build up the organization: enzymes on their own or in complex units in cell particles, mitochondria, and microsomes; structural elements; the cell nucleus with all its content of directing power—the location of the genes of the chromosomal material. All these bodies—bodies that destroy, accumulate energy, regenerate, and direct—are linked to each other according to a functional pattern that gives the cell as a whole a capacity that is beyond that of its components: the ability to reproduce and to preserve continuity through cell division. The functional unit that is a cell is not the result of a summation of the activities of the subcellular units; *it is an integration of organized functions at the enzyme-chemical level.* The whole is greater than its parts.

The cell as a structure is evidently something tangible that can be viewed as a result of the analysis of all the structural elements and other chemical compounds. *The cell as a function* is considerably more difficult to understand, with its turbulent chemical activity involving the continuous destruction and resynthesis of the participating units within the framework of an organization that con-

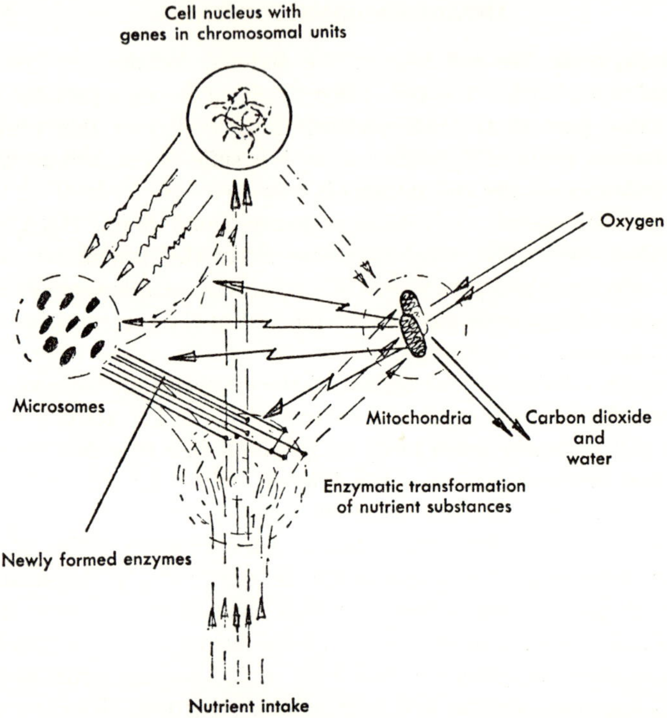

FIG. 9.—A diagram representing the co-operation between the different subunits within a cell. In the genes, which are part of the chromosome structures in the cell nucleus, a certain amount of information is contained that has been inherited through innumerable cell divisions. This information concerns the protein structures of the cell and therefore also its structure and function. The information content of the genes is transferred somehow as a message to the microsomes, which are the centers for the synthesis of proteins, most of which are enzymes. The enzymes synthesized in the microsomes participate in a highly specific fashion in the transformation of food substances. Some of the metabolized food substances are passed on directly to the mitochondria and are completely broken down, with the participation of oxygen, to carbon dioxide and water. This generates a certain amount of energy in the mitochondria, and in the form of active phosphates compounds (ATP), it has an activating influence on new synthesis in general and thus on the activity of the microsomes as well. Some of the newly synthesized material is utilized for a continuous rebuilding of all the units in the cell, including the cell nucleus. Material broken down within the cell passes via the mitochondria into the surrounding milieu

stantly gives direction for its own renewal. We come across the same thing when we begin to analyze the structure and function of the co-operating cell organizations that make up a multicellular organism. Within the whole that is the organism as such, a countless number of specialized cells are organized into a functional pattern. The network of communication between the cell systems is made up of the following types of information transmission:

1. From cell to cell by direct chemical influence by means of a slow diffusion of substances both of nutritional and of destruction products. This plays a vital role in the processes by which a cell—by division into new cells—forms a local organization, for example, an organ or a complex of organs.

2. Migrating cells, which transfer functions from one organization to another, for example, red blood cells and other cells that move constantly. In specialized cases channels have been developed for this transfer of information, for example, blood vessels.

3. Direct chemical action, covering secretion of all kinds and hormonal effects, which is most often transmitted by means of channels of some kind, such as blood and lymph vessels, etc.

4. Chemical-electrical transmission, which covers everything in the category of nerves and nerve impulses, including transmission of rapid directions.

All the multicellular organisms on our earth are variants of this abstract scheme; in different organisms there is a different emphasis in the systems for transmission of information and orders.

What is worth remembering is not the *form* of the co-ordination systems between the cell systems as such—the variations of these forms are in fact unlimited—but the fact that a general co-ordination combined with extreme specialization of certain cell systems gives the organism as a whole properties, as far as function is concerned, that are at a different level than those that could be obtained by a summation of the properties and activities of the par-

---

as carbon dioxide and water. The cell is therefore a highly co-ordinated system in dynamic equilibrium, where structure and function are maintained despite a continuous exchange of compounds. The organization is maintained as long as the food intake and the co-ordinated enzyme reactions function without disturbance.

ticipating units. The biological whole of every multicellular organism is not the result of a summation of the activities of its units; *it is an integration of organized functions.* The whole is greater than its parts.

Let us take the ability to move as an example. If we discount the unicellular organisms that have a certain ability to move in water, cells as such have little motility. An organization of cell units with the ability to contract, plus a rapid transmission of impulses from directing centra in the specialized form of nerve impulses, can form locomotive organs of different kinds capable of rapid movement in water, on the ground surface, and in the air, according to the type and evolutionary history of the species. Their motility is built up from units that have little motility; as a whole they have properties which are greater than a summation of their parts.

From a chemical point of view the development of organized cell systems with specialized functions does *not* imply the introduction of a new factor. The embryonic development of muscle cells and nerve cells and their co-ordination in the developed organism is made up only of a continuous series of local metabolic changes within certain cell groups and a successive emphasis of certain features of the cell metabolism, with certain cell types emphasizing one feature, others another. Nothing new is *introduced* into the system; instead, there is a gradual local distribution of the chemical exaggerations which are latent in every cell, waiting to blossom. When the whole, for example, a mosquito, a mouse, or a man, develops from the division of one original cell of its respective type, this means only that the chemical possibilities that are an integral part of the cells that are formed from the cell of origin have been accentuated according to certain patterns. The same chemical laws hold for the whole as for the cell of origin; the special characteristics of the whole that are beyond the sum of the characteristics of its component cells are the results of a traditional realization of the inherent chemical possibilities of cells to develop in different directions.

Perception develops in multicellular organisms according to the same principles—as a result of an extreme specialization in the development of certain cell regions within the whole. The general sensitivity of cells toward chemical changes in their milieu is reflected in the development of special centra—certain organs for the registration of the chemical fluctuations in the environment—what we call chemoreceptor systems in general, that is, organs of taste and smell in highly developed organisms. Other sense organs have been developed in this way through special accentuation of one feature or another in the general metabolism of cells: organs for the registration of fluctuations in the radiation range between 300 and 3000 mμ—what we call visual organs and thermoceptors. In the same way the pressure-sensitive organs that make up the auditory sense are a special development. In simpler cases the summation of the registration of all kinds of fluctuations in the environment is an incitement to adjustment of the functions of the whole—what we term in more primitive organisms attack or escape reactions before a complex situation. In more complex organisms this registration is more co-ordinated and more localized as a result of a central nervous system that has a more refined capacity to deal with and transmit information from peripheral sense organs and internal regions to effector organs. The power of locomotion of the whole is governed by incoming signals that are modulated by internal conditions. This is not, from a chemical point of view, new; it is only a new form of expression of the potential abilities of the participating cells.

And finally, we can see from this how consciousness develops. To begin with, it is vague with an emphasis on the present; later it develops an increasing awareness of time. The past and the future become realities, and something new has once again been mobilized from the chemical possibilities of the cells. An organization of highly developed and specialized cells has given the whole something more than the sum of its functions, and each function is something more than the sum of the functions of the participating cell components.

This is a very rapid sketch of the development of the functions of perception and consciousness as we see it on earth. It is meant to provide a background for a discussion of whether a more advanced form of life than man exists elsewhere than on the planet Tellus. In itself the problem of consciousness and its development in its different forms is a fascinating one, and it requires special treatment. At this stage we can only become aware of the possibilities that exist for the development of conscious life through other combinations than those that have produced Homo sapiens and glimpse only vaguely some possibilities of the existence of an advanced group consciousness that could be brought about by a special form of organization of units with an individual consciousness.

Let us consider the first aspect. The basis of human consciousness is the following:

1. Possibility of contact with the outside world through (*a*) incoming signals from sense-organs registering the fluctuations in the milieu, including signals from fellow-organisms; and (*b*) outgoing impulses, locomotive power with an ability to interfere with the milieu, including signals to fellow-organisms.
2. Internal processing of data from the surroundings and from the organism itself, leading to modulation of the outgoing signals with respect to intensity and quality.
3. Integration of the neuroelectric and the chemical networks into one unit in which the function of the data-processing central unit depends on the finely regulated co-operation between those units that produce nourishment and those that provide special activators. This is in turn influenced by the central unit for activation of the central-activating activity, etc.

This all takes place within one and the same architectural unit, man's body, by a local play of afferent, efferent, and the remaining other impulses. These impulses are activated by the intake of simple food substances by the participating cells and are expressed as judgments and initiatives that exist in the central unit as consciousness, ego, and id.

Is this the only solution to the problem?

In his excellent futuristic novel *Last and First Man* (1930), Olaf Stapledon discusses a theory about the evolution of consciousness in a collection of biological components that implies that the contact between the units is not based on the transference of chemical or electrical information via material channels but on short-wave messages of some kind between physiologically independent units, which he calls virus types. He draws an interesting picture of a fairly diffuse mass of complex units, all of which are chemically specialized but which are not bound together into any morphological structure within the whole. The integration of the units into the whole is achieved by an exchange of information within the whole rather than through a limited system of chemical exchanges by means of fixed duct systems, and in this way a kind of biological collectivism is formed that has every possibility of creating *group consciousness* in accordance with points 1, 2, and 3 above.

Stapledon, in short, argues that the development of conscious clouds of particles on Mars took place as a result of the difficulties of the local life in a state of ever increasing drought and shortage of oxygen. The units he envisages are endowed with the capacity for photosynthesis, are submicroscopic, and have certain features which—as we can see thirty years later—are reminiscent of chloroplasts in green plant cells or photosynthetically active mitochondria, which we will probably find are two aspects of the same thing. These units, stirred up in a thin and dry atmosphere, individually have little capacity to communicate with other similar units near them. Several units, however, meet and function in close contact with each other, thereby forming a unit with an increased ability to communicate. Many small clouds, organized according to a certain pattern, are able to emit electric impulses of a wave nature; this is conceivable in terms of a development of the theme of electric fishes on *our* earth. Many small clouds organized into a unit become a transmitter. One central transmitter directs the organization of certain tuned-in units into larger units with special chemical or mechanical tasks. The solutions of these tasks are relayed to the central unit, which considers further directions, and so on. The whole system is based on the principle of the complete chemical identity of the subunits but also on the functional dissimilarity complexes of these units that are formed through the establishment of a system. Information is going *inward* toward the center and orders for the activation of existing organizations, for reorganization, or for concentration of activity in a certain direction, and so on, are emanating from the center. It would comprise a form of consciousness based on the organization of identities.

According to Stapledon's theory, activity within this type of organization is relatively independent of the positions and distances of the subunits. On the other hand, the activity of the whole is dependent upon a definite concentration of units in a small area, and this process of contraction leads to the formation of something that is *capable du tout*. It is this black, viscous something that becomes hostile when confronted with some of the species Homo sapiens who are equipped with interference transmitters. The complications become quite embarrassing for both parties. All according to Stapledon.

A variant of Stapledon's theory of collective consciousness is found in Fred Hoyle's *The Black Cloud*. In this he describes a number of units, forming a rather formidable intelligence, that are held together by the exchange of information via ultrashort-wave transmission. A rather unusual feature that Hoyle introduces is a supply of material via electromagnetic direction of gas currents within the cloud; but this could be plausible under certain conditions and from a new point of view on the subject. But up to now we have not seen anything that is at all similar to this on earth.

Both Stapledon's and Hoyle's theories of the development of consciousness through subtle organization of organic subunits are not inconceivable in the way they have been formulated by their respective authors. Both systems resemble in some ways what we interpret as group organization in insects. We have here a small number of types, each of which has a countless multitude of hereditary patterns that make up the functional units within the framework of a society; the *organizational pattern* of the society exists and survives even if the participating units disappear and are renewed annually. This is, in other words, another version of what happens in every living cell; the organization persists—for some time—the material is simultaneously broken down and resynthesized while crude material, food, is being supplied, in certain cases with the simultaneous addition of energy.

The parallel with ant or termite societies is, however, only valid from a certain point of view. For these societies are a striking example of group *activity,* not group *intelligence.* What we see as an organization with different forms of social expressions is the result of a hormone-regulated automatism; the ant or termite queen influences the behavioral pattern of a number of specialized units by chemical means, so that the work of the individuals results in a society that protects the central unit with all its unconscious executive power. The society is not a product of consciousness at work but is an automatic result of a chemical organization at a considerably lower level. It is somewhat similar to the colony formation of fungal spores or of embryonic development in its preliminary stages.

A more accurate comparison to explain what Stapledon and Hoyle really mean by their proposed models of consciousness that have no morphological organization is a research group that has no actual director—a group in which experiences are shared by the group as a whole and in which the summation of the team work and the integration of its results influence its future activity. The lines of communication that exist on earth today for the issuing of orders and the exchange of information could lead to an integration of human consciousness, according to the patterns that hold true for the human body as a whole, by means of which the results of group activity could become greater than that of its participating units.

We must issue another warning about drawing too many conclusions from the ant society, with its rather obvious analogies for the human social level. What we are thinking in terms of is the development of a group intelligence, not through *summation,* but through *integration.*

At this level we can imagine that conscious life could exist in space. The difficult thing is to imagine how such a superorganization would "think." We can well imagine that the information system could be a well-co-ordinated network of signal channels from different centers that perceive different things; this would be similar to all analytical work within all branches of science and research. The outgoing orders pass through a similar distribution network. So far, so good. But how is the data material treated in the body that is not *one* individual—from our viewpoint—but a multitude?

(A parenthesis: How do *we* think? We know how the incoming signal system works and what the nature of the incoming signals is; we can see in the outgoing signals—speech, writing, actions, orders—the results of the processing of the incoming and stored data. But how was this processing achieved? How can anyone analyze another person's line of thought in all its chemical-electrical complexity?)

What we find so difficult to understand is that the complexity of the data-processing system—that is, our own brain—is too great for us to comprehend with our senses. We may know in principle the ways in which the machine works; we can conceive the incoming and outgoing signals; but we find *the pattern of operations that leads from the incoming and accumulated signals to a conclusion too complex for us to comprehend.* The truth is that we can only understand and analyze a small number of processes at any one time, whereas a conscious thought contains a couple of million microprocesses co-ordinated into a counterpoint too advanced to be captured as a concrete score.

But our brain functions despite the fact that it works above our—and its own—ability to understand its functional pattern in detail. In the same way, we can imagine that a superorganism—formed through the skilful co-ordination of conscious subunits—could solve problems of formidable dimensions; but the *pathway* to the solution of a problem through the processing of its own impulses and those relayed by myriads of units would present the superorganism with some difficult problems, as it would the single units that would have put their simple consciousnesses to the service of the organization. The more such an organization is refined, the more self-repairing it becomes; the more autonomous it becomes as far as the utilization of local resources of power and material is concerned, the more unsurveyable becomes its functional pattern by which the incoming data, questions, and problem complexes are processed into outgoing data, answers, and complex orders.

We are only beginning to realize this; but in many places in space it may be considered as elementary or quite obvious. The guiding principle in the developmental endeavor of an advanced organism might be that the phenomenon that started in many places in space as a kind of game with the inherent reactivity of the carbon atoms as a stake and finally became planetary life and consciousness will always become gradually aware of its own limitations: that *consciousness will always be one dimension above comprehensibility*. Despite this, it is important to search for new methods to try and diminish this distance between comprehensibility and consciousness to a minimum. In striving for this goal such simple things as space travel and intergalactic communication will gradually become an aid—one of many—to get closer to something that will probably grow more difficult to understand the closer we get.

Most advanced organisms have probably already discovered what the most favorable shape is for their own organization, according to what the local environment can offer in terms of material and power sources. There are countless alternatives, the realization of which on a practical scale, biological as well as technical, would probably confuse a traveler from a less advanced planet. The well co-ordinated clouds of global dimensions which we have previously discussed are one solution. Explorers would probably not notice them at first, until some of the local groups of units have contracted into a more palpable shape in order to give a friendly welcome, or perhaps the opposite. Another global solution of the integration problem could be the existence of organisms in a network around a planetary surface, forming a kind of superorganized nervous system with certain connected flexible centers by which refined muscular effects ranging from rough work to micromanipulation could be achieved. These are systems in which it would be difficult to localize a real center of activity—it is difficult to do this in our own brain—and which would probably seem uninteresting and silently inactive to an imperceptive visitor, if he noticed it at all.

Units that can move are probably just as common, however, and these would offer certain advantages to the whole as far as a satisfactory co-ordination of the units is concerned. An immense number of systems could be built up; a permanent radio connection between individuals and a central data machinery would be the most practicable solution. It is probable that we will find sometime that a certain specialized development of individual consciousness—at a level far above ours—is a practical solution for the mobile units and that the processing of problems above this level will be left to mechanical equipment that is either of a conventional electronic form or of advanced bioelectric variants. It is possible that this development—with all its possibilities of producing a collective brain on a large and strange scale—would have the sensitivity and flexibility to process incoming signals from myriads of superintelligent units and to formulate superintelligent orders at an even higher level.

A feature common to all forms of advanced consciousness will in these circumstances be that each separate unit, either in a cloud, in a network, or freely moving around, will always be limited by his personal ability to survey a situation, even if his capacity for data-processing is far beyond that of any individual on earth at the moment. Our own global organization of human intelligence is clearly still in its first developmental phase, even if projects for the achievement of certain goals are beginning to assume a more definite shape. Projects that are often in complete opposition to each other are often pursued through lack of contact between the projects, a state of affairs which may only be temporary but which would probably shock a more advanced visitor, that is, if he considered us worthy of a closer inspection. By a return visit sometime in the future, we could gain—by an exchange of ideas at the individual level—knowledge of how far technical developments—both theirs and ours—had taken us toward an advanced form of consciousness. An encounter in the future. Somewhere, sometime, light-years away from our home borders.

The members of the crew of the primitive spaceship from the planet Tellus slowly get out of their meteorite-scratched ship and step for the first time on firm ground on the planet $3,\gamma^2$-Eridani. They have spent 301 years in traveling, in a frozen state over the longest part of the journey. A planet of the size of the earth has given them something to look at during the last few days of a long journey. The landing has taken place according to the instructions exchanged between the two planets, and now the explorers see a foreign sun shining over a landscape which has no great contrasts and where a few persons are working—it is difficult to tell what they are doing in the sun haze. A strange looking vehicle drives up at speed bringing what seems to be a reception committee the members of which are of a somewhat unusual shape although they nevertheless exhibit a certain individuality. A mechanism delivers a greeting in the code language developed for the communication between Tellus and planet 3. Everything appears to be friendly, and so there is some cautious fraternization. One can only talk with the aid of code equipment, and a member of the crew asks a somewhat unreal-looking being who seems to be especially interested for some reason: "Are you really so advanced in all things? You signaled long before we did." The green-glittering figure looks at the questioner with melancholy eyes and answers according to the usual syntax: "I am an astrobiologist from X. I don't know the project in any detail. My job is to analyze you before you go to H. You must stay here. It is an order from Big Brain. What do you bring?" The questioner turns to one of his fellow crew members and mutters: "I have a feeling that I have met this guy before somewhere; I can't think where. I wonder if he really represents anything more advanced?"

CHAPTER XI

THE LONG-TERM AIM

A guiding principle in the plans that are made for the future is the development of a supertechnical social automatism that will leave few opportunities for individual development on the part of the participating units. The classic conflict between the individual and society will become even more pronounced, and the future for any individual who for some reason or other finds himself opposed to the social politics that have been accepted for the region in question is a gloomy one. This theme crops up again and again in the sketches that are drawn of society in 1984 and later on of the brave new world and the outposts of the future that have been visited by lonely investigators with access to a time machine. In all of these worlds advanced technology is used to bring pressure to bear on the individual and his behavioral pattern, either by methods which are highly sadistic, by a superorganized benevolence that is concerned primarily with the stability of society as a whole, or by the development of a completely automatic dependence system under the token of an organized unconsciousness. It is difficult to know which is the worst.

To most authors who develop the idea of an advanced technocratic society its final outcome appears to be an unmitigated hell, and they feel that our own time will in retrospect appear pleasant and worthy of nostalgia. Very few writers have had any positive views in their analysis of what man's future existence will be like during the next couple thousand years. Those who for some reason feel optimistic about the future usually, on closer examination, feel that the present is another version of that unmitigated hell. Escape into the future is the only way out of a situation that is so unbearable that nothing could be worse. So their expectations of a future utopia are usually rather modest, and they often have a "back to nature' quality about them.

It is interesting to compare our attitude toward a technical society in which man would be confronted with man in the distant future, with the corresponding situation of an encounter between man and representatives of a foreign biological development. In the latter case man would be little discontented with his own advanced technical development. Access to all the paraphernalia of advanced variants of atomic power and improbable biological knowledge in a well-developed technocratical society would be considered completely natural. In addition it would provide an extremely valuable support in those situations where man is confronted with invariably hostile visitations from another part of the universe. In this kind of vision into the future, man develops his best qualities as a defense for his superorganized home environment, without letting an enormous potential of technical and scientific knowledge influence the good points of his behavioral pattern in any way; friendly, helpful, courageous and chivalrous, he gives the impression of having been able to see the development of technology just as an interesting phenomenon. He refuses, however, to assume an inhuman attitude in his capacity of participant in a new development. Buck Rogers and his colleagues, bless them.

If we eliminate the more naïve features of our attempted extrapolations into the future, some general tendencies will remain which will express themselves—both on earth and in other regions of the cosmos—in certain definite developmental patterns at both the technical and the biological levels. It will be like a play that is put on in different versions, but with the same moments of tension, by a differing stage management, aeon after aeon.

Let us consider, to begin with, the time factor. From the point of view of the earth, the energy production capacity of our sun is a limiting factor for the future evolution of life within our planetary system. Calculations of the present energy production, based on the catalytic transformation of hydrogen into helium, show that forty billion years is a realistic estimate for the time that the sun will remain in its present shape with approximately the same intensity of radiation.

In this case, life on our planet has a future in relatively unchanged external conditions that corresponds to eight times the whole period of time that has passed since the formation of the planet and approximately fifty times the period during which organisms of a modern shape have existed at all. This situation seems quite encouraging from the point of view of time, even if we allow for the possibility that some cosmic calamity may occur during the next forty billion years. The continuous transformation of hydrogen into helium is accompanied by a successive change in the dynamic equilibrium of the condition. At a certain ratio of hydrogen to helium the solar condition becomes labile and, according to current theories, a new cyclic process might suddenly start. The energy production could rise to at least ten times the present level, after which a rapid and continuous decrease in luminosity would take place, finally reaching a level far below the present one. This temporary flare-up might become a permanent reality before the forty-billion-year period is up. The threat will probably become acute in ten billion years, or perhaps even earlier.

The results of this solar explosion, which will probably take place without very much warning, will mean that temperatures far beyond the Mars orbit will rise to a level that will make any life at all impossible. All vital activity that may exist in the region from Venus to Jupiter will be extinguished in the course of a few earth years, perhaps even more rapidly. On the other hand, the outposts erected for one reason or another on one of the satellites of Saturn, Uranus, or Neptune will suddenly have a pleasant improvement in their climate. To the extent that it is possible technically, emigration on a limited scale to these distant places will, of course, be carried out; but man's attempts to start all over again and build up a new civilization in this completely lifeless environment will be an almost incomprehensibly difficult problem. Moreover, the colonizers will have to take into account the fact that they will have to move back toward the center as the radiation of the sun gradually decreases after its first cataclysm. Since this latter phase may extend over some ten billion years, it seems long enough, from our point of view, to permit preparations before every new step; but the whole situation has a touch of melancholy, an awareness of cosmic rootlessness and of the continuous necessity to prepare for departure.

It is most probable that this situation is already an acute reality for many systems of conscious life in the universe. There is probably a movement outward in space prompted by the threat of a flare-up by the central sun or suns in many planetary communities within our galaxy. Or perhaps there is a corresponding migration toward a shrunken sun, with its main energy spent in a few years of nuclear exuberance. But whatever is happening, there is one feature which is common to all forms of conscious life—the automatic mobilization of all the possibilities of survival. Movement in space within the local planetary system would be quite normal for a highly developed conscious form of life in order to secure a continuity for its development. It would not be only a thrilling achievement, as it is to us with our time scale of one or ten years; it would be part of a long-term plan based on an entirely different time scale.

One or two billion years is a sufficient period of time, however, for conscious life to evolve, irrespective of its shape and environment. The most critical period in this evolution—we assume that the external conditions are fairly stable—is the first phase of development, when industrialization and the use of atomic power become widespread. During this period, if some fundamental errors are made, they could lead to an over-all destruction of local life in general, possibly even to its complete annihilation; they would certainly lead to a serious interference with development. After this difficult phase has been passed safely, a slow maturation process would probably set in, varying in its local forms in ways which would seem quite strange to us now and probably frighteningly different from our present conception of the formation of a society.

Irrespective of what the forms are, either at the individual or the society level, advanced systems will have some common technical problems. The energy supply can be based on atomic power to a large extent, but this will require a supply of raw material. The natural concentrates that are present in the upper strata of the ground surface will probably be consumed within a short period. The rich ore deposits of the earth are a geologic rarity rather than a rule, and we can assume that they will be consumed within the next ten thousand years. A civilization at an advanced stage will probably have to work with extremely poor material in order to produce certain essential elements. Even with highly developed enrichment techniques and a reliance on the use of atomic energy, such metals as copper and silver will gradually become scarce and expensive. Aluminum will dominate the market, and plastic carbon compounds will be used more and more widely as a construction material. All this will be an inevitable result of a common factor, that is, life on a planet. We can assume, as we have pointed out earlier, that every planet in the universe has the same atomic constituents of its core.

Phenomena that we term technical developments and industry, with their utilization of either local or extraplanetary energy sources, will always remain of secondary importance to the central problem that every advanced organism will have to solve: its nutritional requirements. In whatever ways the environment is shaped, the transformation of its materials to meet the constant demand for food will remain a problem that will require continuous attention and development. Biological research and advanced biotechnical work will become a necessity for the development of any advanced form of organisms or society. *How* this development at a biotechnical level will take place in different localities of the universe is, of course, unpredictable from our point of view far away, but there are certain lines of thought that are worthwhile following up.

As we have pointed out earlier, the molecular organization of living organisms, wherever they are encountered, ought to have certain features in common. For example, carbon compounds of the protein type, carbohydrates, fatty acids, and certain cyclic carbon-nitrogen compounds will probably exist in the mechanism for the transformation of material that is present in all organisms. From this point of view, locally occurring carbon compound material is the basic element in the process of maintenance. Carbonates and carbon dioxide or local hydrocarbon compounds in combination with water, nitrogen, and phosphorus compounds become the representatives of the inorganic milieu from which—through the influence of energy—the local organisms carry out the intricate transformations required to build up their own material in order to exist in the dynamic state of equilibrium that is life, regardless of the milieu. At the same time some material must return all the time to the level of the simplified and unreactive something that is the non-living milieu.

In whatever ways, however, this general theme has been worked out on different planets, there is probably one common factor: the utilization of the radiation of the local central sun (or suns) as an energy source. On our earth our special form of life has been formed as a result of photoactivated cleavage of water into oxygen and hydrogen. A special combination has allowed the intake of slow-reacting carbon dioxide to form complicated plant material, with the food intake of other organisms carrying out a clearly antagonistic reaction with carbon dioxide and water as the end product. The important point is that photosynthesis *requires space*. Any intelligent organism will realize that there is a limit for the synthesis of organic material per area at a certain intensity of incoming radiation, with the upper limit at no more than 4 $cal/cm^2/min$ for cultivation in the open. At higher intensities it is simply too hot for complicated organic material to exist. As a result a certain relationship must be maintained—as on the earth—between the cultivated area and the number of leader organisms. In extreme cases the latter require the maximal utilization of the whole planet surface for controlled cultivation. This is an extremely artificial situation that utilizes every unit of space in order to maintain a balance between availability and consumption of food stuffs for a population that has grown to the maximum number possible to maintain.

We may assume that the development of such a situation implies, at the same time, a successive elimination of all the unnecessary intermediate stages between cultivated organisms and consuming superorganisms. This will lead to a planetary milieu with extensive areas of cultivated organisms of a limited number of types and an otherwise drastic reduction of the flora and fauna to a minimal number of species that have some utilitarian function. Such an evolutional development is quite possible, and it will automatically make advanced biological research and long-term planning essential. It is most probable that knowledge gained at the general biochemical level will gradually bring about certain conscious changes in the types of leader organisms themselves by methods that are far beyond our present methods of plant improvement and eugenics.

Within the framework of such a project many variations are, of course, possible. The two possible extremes can be represented, on the one hand, by an attempt to achieve maximal individual versatility within a homogeneous population and, on the other hand, by extreme specialization and a great number of types at the individual level within an extremely stable social structure. Prototypes for both of these developments in man have been advanced by writers. Olaf Stapledon, for example, hypothesizes the future existence of the fifth man with an average life span of several thousand years and with a pronounced physical and psychic stability. He sees him as a massive organism 3 m tall, with extreme versatility, especially in manual work, with a thumb that can grip with a power of several thousand kilograms and a little finger with the sensitivity of a micromanipulator. He will have an extremely powerful brain, which will be able to solve problems of appreciable dimensions, and he will possess a highly developed consciousness both as an individual and as part of a group; he will be the product of several million years of industrious biochemical-genetical research, which has resulted in a uniformity in form and a versatility in functional abilities. Aldous Huxley draws the opposite picture, with man in the future existing as a multitude of type units, the co-operation of which through an extremely well co-ordinated behavioral pattern at the individual level gives the society as a whole a high degree of stability and a possibility of group consciousness that is concentrated into a hierarchy of leader organisms. These two illustrations of the results of evolutionary tendencies may seem somewhat absurd, but they are both within the limits of what is biologically possible.

Regardless of its individual or social form every organization of conscious organisms will gradually discover the optimal conditions for its preservation and maintenance by an increasingly intense utilization of the local and extraplanetary energy sources. The earlier phase of every such evolution will probably be characterized by the utilization of photosynthesis, not necessarily in the sense or in the form that it takes on the earth, but similar in essence: the activation of slow-reacting carbon material into a higher degree of complexity, represented by primitive organism material for consumption purposes. In the first place solar energy will be utilized as an energy source, but later on atomic energy in different forms will become the main source. After thousands of years of fluctuation and instability, the prerequisites should exist for a social structure so stable that a development toward advanced thinking will not be threatened by the failure of short-term projects aimed at establishing control over problems of maintenance and overpopulation—things that are disturbing features in our own primitive system just now.

The prognosis for the relatively undisturbed evolution of intelligence influenced by humanity—these symbols are our own, and other forms of life will have other variants—seems to be clearly favorable for any consciousness that already has several million years' evolution of mental activity. How long such an undisturbed period of progressive work can last is another matter. There will probably be some interstellar differences of opinion that will cause local anxiety; there might even be some more serious long-term problems. For all life on any planet that is dependent upon a central sun, there will always be an element of acute insecurity—that is, how long will the sun last?

For many units this problem will seem too remote to prompt any active discussion about what to do in case an evacuation becomes necessary. For other units in the universe this problem will already be an immediate one and will demand carefully prepared or rapidly improvised solutions of the formidable problems that will be posed by the threat of a possible sudden explosion of the central sun.

The closer a living organization of consciousness approaches the critical period for its central stellar body, the more pressing the question of the departure and expansion of the essential parts of the organization will become. This will demand intellectual cooperation to a far greater extent than during the early periods of local expansion on a small planet. Early settlements on planets on the edge of the system, which because of their excessive cold or isolation are difficult to colonize, will gradually become something more than an expression of technical developments; they will become a necessity in the future, which will require long-term planning and organization so that an evacuation could take place at short notice. On the whole the measures that these different organizations of life can take are the same regardless of their social structure and individual forms. Preparatory work should be carried out on a distant outpost to make the milieu more favorable for the transfer of life that has up to then flourished on planets with more agreeable conditions.

The high level of biological knowledge that has emerged earlier from local expansion made necessary by excessive overcrowding and acute lack of food will be faced with new problems of a complexity difficult to realize. Awareness of the brevity and artificiality of the milieu that has been built up by any advanced organisms will create new possibilities for mental acumen and inventiveness. It might also create a greater understanding of this problem from a universal point of view It has happened before, it is happening now, and it will happen again in the future; our own situation is not unique.

The measures that the population of a centrally threatened planet could take would naturally depend on the local possibilities of emigration. Generally speaking, wherever life is present in a more advanced form, it is probably to be found in relative proximity to a central sun, or suns, in most planetary systems. If the unstable sun is a G sun, like our own, the chances of life are greatest at a distance of 100 to 200 million km from the center. There will probably be one or more outer planets with arctic climates that will be available as temporary retreats until the first cataclysm is over and a move back toward the center of the system is possible, if, of course, there is anything left to move back to. In this situation the first step would probably be to transfer a frozen concentrate of local life to an outer planet with the intention of activating this implant when the radiation increased suddenly on a frightening scale following the eruption of the distant sun.

Such a plan is by no means impossible. On one of the larger Jupiter satellites, J. IV or J. V, for example, certain specialized micro-organisms could utilize thawed water and atmospheric methane and carbon dioxide in order to reproduce, even after a long frozen sleep. We could select at this moment micro-organisms with a capacity for photosynthesis in the presence of hydrocarbons. The full development of such a microlife could give good biological support to the few pioneers who would be transported to these distant regions by conventional spaceships in order to start the colonization of what has earlier been an outpost, an eye toward outer emptiness.

This would only be a small part of the plan. An important factor in the preparatory work would be, of course, the successive shrinking of the population, in different ways according to the locality and the social structure, either rapidly or slowly; however, this problem is beyond our investigation. Finally, a small number of people who have been selected over a number of years —centuries or millenia—to continue the species in another place will depart from the home planet. This whole operation, which seems like a tragic evacuation, will occur so slowly in most places that the suddenness of the crucial decisions will not be realized by the species for a long time to come. Under these conditions the future departure will probably be looked upon hopefully as providing a new home in another location in close proximity.

But what will it be like for those who do not have any planetary neighborhood to move to or who have already moved to the outskirts and are faced with the necessity of struggling, step by step, closer to the center of a dwarf star after the big explosion? They will be faced with a transfer program of cosmic dimensions, with a desperate attempt to leave a local planetary world, threatened first by destruction by heat and later by excessive cold beyond the limits of the endurable. There appear to be three solutions to this problem that are within the limits of what is technically possible.

One of these is to launch, in a desperate insemination attempt, shipments of the essential parts of the species in the hopes that one or a few will land on a planet in a nearby solar system and develop in undisturbed peace into a primitive form of local life, from which eventually the thoughts of the sending body can again arise on foreign ground and continue to live. This form of life would probably remain unaware of earlier events and difficulties in a naïve unconsciousness of anything else than the immediate evolution. At the present we have few chances of carrying through such a program ourselves; but it is not too difficult to imagine a technical development some time in the future that would permit a concentration of genetic factors within a cellular frame that would contain a concentrate of what we consider as our own biological characteristics. After its journey to distant worlds, it could be aroused to activity, perhaps to active consciousness, perhaps within a relatively short period in a suitable chemical milieu—after hundreds or thousands of years in frozen oblivion of the past. It would be unaware of every past initiative, of everything; and yet it would be a continuation.

Another alternative would be to send a number of people in spaceships as a concentrate of surviving life at an advanced level. They would travel in the direction from which signals indicating the presence of conscious life had been received. There might be no signals to follow, in which case they would go not knowing where they were going or what they would find. It would be a gamble on a supreme level, with the survival or absolute extinction of a species as the stakes. It would be an advantage from the point of view of the conservation of food for the personnel to be frozen during the long journey. To awaken in a friendly and understanding milieu after this long journey would seem like a miracle. It certainly would.

A final alternative, and the only thing to do if the other alternatives failed, would be to stay in the cold or burned-out place and wait and see and try to make the best out of a hopeless situation, with the full knowledge of the inevitable end of life on the planet. It is probable that in this most isolated of situations, with individuals and groups beyond reach of rescue expeditions from humane fellow organisms on more favored planets, lethargy would slowly set in, despite the possibilities of creating temporary relief out of the debris of a burned-out planet or sun with the aid of atomic energy. Some individuals might in this situation contemplate—in an ironic mood—the mercilessness of the universal power that had permitted the intrinsic power of matter to manifest itself as life for only a short period of time. It is possible that in this situation thought will finally reach the level that it has been seeking for billions of years. What the answer will be is known only to the one who understands all. It is an answer that cannot be transmitted as a simple signal through great space, before the evening closes in.

The universe that we can survey at the present time holds some 10^{10} stellar systems and galaxies, each with an average of 10 billion suns; thus there is a total of some 10^{20} units of excessive activity, the manifestation of nuclear energy dissipated on a large scale. Some of these units are of the same dimensions as our own sun, both as far as size and energy production are concerned. Some of these suns—and probably many others—have planetary systems. On a conservative estimate, based on the possibility that one out of a thousand has material bodies moving round it in orbit, there are at least one billion planet systems. Some of these contain—as part of their present, past, or future structure—a planet in such a position relative to the central sun that the incoming energy falls between the limits 0.5 to 4 cal/cm^2/min, according to our measurements. There is at least one of these planetary units in every thousand planetary systems; so there are at least one million. On some of these life has developed. One of these planets is our own.

A sun, type G9 flares up. Warning signs have been visible for several thousand years, and steps have been taken on nearby planets, which by a strange coincidence have also been endowed with conscious life. Signals have been sent to adjacent solar systems, but there has been little response, except from a far distant unit that would not be able to participate for the next five hundred years. The evacuation of the threatened home planet proceeds according to plan—for once there has been unanimity about the goal but not, of course, about the methods to achieve it—and a dwindling population listens with satisfaction to the reports about the increasing percentage of hits by biogenic units on planet 17, which up to now has been thought of as an interesting observation point but which now has become a goal. Time passes. One lives and dies in the artificial world that has been created with so much effort; one tries to live as usual, but . . . there is a feeling of departure in the air. We have known this for a long time. It all started when we first saw the pulsations of the sun and watched the intensity of the protuberances increase and die away in a rhythmic variation. We realized that our existence here at home was drawing to an end; everything that we had built with so much concern and work was doomed; our future lay in expansion into space and to other planets, where alone we can continue in ways that we feel to be typical of our own individuality.

It is difficult to understand this sudden change from a stability that has been achieved through an understanding of life built up after billions of solar years to this extreme rootlessness in which a distant planet, where everything will have to be built anew, is the only hope. . . . On the other hand it holds out a promise of an adventure, and we need this after aeons of stable existence. But it has come so quickly, we are not yet prepared either as individuals or as a group. We cannot understand our individual situation or the meaning of it all. We have to go; we shall have a lot to talk about when we meet again. . . . We are not alone in the universe.

A sun, type G9, flares up. One tradition is annihilated; another continues.

CHAPTER XII

THE TASK

Our thoughts have expanded from a very small world toward the limits of the universe in time and space. We have met with some local and some universal problems; we have become familiar with our planetary system; and we have glimpsed the structure and evolution of a larger whole. The picture we have is still diffuse and shadowy, but we know that in time its outlines and colors will be drawn in more exactly, if we have the time and the opportunity to develop our capacities to comprehend and feel.

From a universal point of view, time seems to have been allotted so generously as to allow the evolution of a newly awakened consciousness. We have seen that the prognosis for the continued existence of our sun as an energy-giver gives us a fairly undisturbed future of some ten billion years, perhaps even four times this figure, with the reservation that a spontaneous outburst is always possible. But even so we should have a good chance of enjoying an existence within our planetary system that—as far as man is concerned—represents a span of conscious life one thousand or ten thousand times longer than the past ten thousand years of dawning perception. Ten to one hundred million years in which to develop and expand our possibilities, starting from now.

It seems to be a long time, but we may need it in order to carry out a task. What task?

Many organisms all around the universe are probably asking this question at this very minute—primitive beings who have just become aware of the problem as well as those advanced forms of consciousness that have been thinking over this problem for a few or many million years and have still not arrived at a final solution.

For all organisms, either individuals or groups, primitive or advanced, there is a starting point that has certain common denominators. All those who have passed the first stage of perception of the structure of matter and the form of the cosmos must inevitably have found out certain elementary facts about life and its chemical background. They must also have realized the universality of the chemical possibilities. They should have realized at an early stage, just as we do now, that the distribution of chemical elements is about the same all over the cosmos and that all life, regardless of its façade, operates according to certain common factors. After a time they will realize that the chances that they are unique in the universe are small. Probably some have already established contact with each other either by means of long-distance exchange of information or through visits to each other's planets. Others are biding their time and are exploring their own planetary systems. Some may prefer to stay at home and continue their isolated existence without contact with the growing community.

All forms of conscious life scattered throughout an expanding cosmos will sooner or later be faced with this question, Is there any specific job that we can do, regardless of time and space, now that we are aware of the common features of life? For inexperienced organisms with only a short time of expansion behind them, this question brings up many supplementary ones: We now know that we have a long period of existence before us and that our knowledge of material, life, time, and distance will increase considerably, but will this development be merely automatic? Will we be unable by our own achievements to shape the general trends? Will we achieve anything, universally speaking, by organizing our consciousness to learn more, understand more, within the short time we exist in our own world? Can future contact with other inquirers help us, or is the answer within ourselves? Is the evolution of the whole something more than that of its components?

By asking these questions—which are more than mere requests for information—we begin to suspect that the final answers to our questions are far beyond our power of apprehension and that, anyway, to obtain answers would not satisfy our deepest needs. The more we consider what we are really asking, the more we hesitate to be confronted with the answers, as we could imagine them to be formulated by something with a universal experience.

. . . By now you must have begun to realize that the developments of which you seem so naïvely pleased are an expression of the automatic feedback of your attempts to solve the basic problems concerning matter and its inherent power patterns. Technical developments of simple processes, with subsequent utilization of atomic energy and expansion into the neighborhood, are things that always flare up here and there after a cosmic microsecond of widened thinking. It is inevitable; it is an intermediate stage in the embryonic development of consciousness, a means more than a goal. An attempt to accelerate the technical aspects of this process is neither a plus nor a minus accomplishment, for the tendency is inevitable despite all the implications of conflict during the earlier phase. In general your evaluation of what is happening in terms of plus and minus is a primitive method of self-assertion, which will be modified later when you acquire new means of expression through learning more, understanding more, and above all, feeling more. The latter aspect is fundamentally important from a universal point of view; the others are only aids, which, to the extent that they are utilized for the further development of sensitivity in the individual and group consciousness, have a plus quality. No contact with anyone can guide you in the development of this quality. We cannot formulate for ourselves what we would like to express. However, the following might give you the idea in general terms.

In its evolutionary tendency the universe is a manifestation of great power and variety; it is a play between the subatomic particles in the dynamic system of the suns, between particles in interstellar matter, both larger and smaller. The inherent tendencies of the basic material to form more complex units, especially the formation of molecules, are expressed, particularly in the few regions that have a suitable yet a narrow temperature range, not only as a crystallized result but as a continuous local combination process. In rare cases certain carbon material may form complicated structures, with sensitivity toward incoming light energy. As a secondary development of this phenomenon, which at the chemical level implies a profound activation of the combination possibilities of carbon compounds, an evolution of certain systems will eventually occur. We can call them organisms, and they are endowed with a certain sensitivity to time, distance, and structure, and therefore to influences in general from local phenomena and from the universe as a whole. There is another secondary development of this phenomenon about which we are not yet clear.

But every perception of anything outside the local milieu is an incitement to development. Of course it does not have to be; in countless cases where photosynthesis, such as in plants, has begun to develop, the local conditions or the cosmic events combine to create a milieu which is unfavorable for further development. From an over-all point of view the perception of light energy that begins as photosynthesis in plants and later becomes a phase in the evolution of the eyes of animals—and your own eyes, for the matter—is a stage of evolution, where the unconscious perception radically changes a developmental tendency that is completely bound to the locality. The evolution of photosensitivity on your earth has had an activating influence on the chemical processes all around your globe and has led gradually to the evolution of systems with a pronounced ability to integrate light impressions as well as other signals from the surroundings. In a short time organized units were formed that had the ability to glimpse both local and universal objects and the ability to perceive impressions from a whole, including signals from outer space. The results of this are influencing your evolution toward something new, something that is as different from your present level as that level is from the time when photosynthesis in its modern form was a novelty.

This evolution of a new quality that is beginning within yourselves, the ability to integrate impressions of structure, changes, and events, both in yourselves and in the universe that you can survey at the present time, is something that requires more than knowledge and comprehension. Just as an instrument can give better results the more highly developed its sensitivity is, to use an ordinary example, so can an organism contribute more as an instrument of understanding, either as an individual or as a group, the greater its sensitivity is. So we need to comprehend the evolution of the universe as well as the events that led to our own development as an instrument. To develop this sensitivity is a job that is not altogether pleasant, either from a local or from a cosmic point of view. But it ought to be done.

Why? From a universal point of view it should perhaps be done in order to understand some of the greatness of the whole universe, before the expansion of the systems within it has pushed the conscious units so far apart that contacts between galaxies and worlds become too difficult for purely technical reasons. Exchange of information in the future will always have an undertone, however, of the incomprehensibility and beauty of the universe in its structure and dynamics—a realization of a cosmic adventure. Sometimes the response will be at another level: sensitivity in the apprehension of local phenomena, keenness as to nuances, appreciation of differences, pleasure in overcoming difficulties.

And this is the end. The question whether the evolution of the whole is of greater importance than that of its components is not very well formulated, as you may now realize. Obviously, the evolution of the separate components is of greater importance than that of the whole, as this is the only way in which the whole will get an opportunity to understand itself. This is how we participate in the whole. This is where our job lies. . . .

Two units look up toward a dark night sky with its myriads of solar systems, galaxies, and cosmic gravel. Every second the distance between them increases by 50,000 km in earth measurements. Two worlds that are a distance of 1.3 billion light-years apart. Each sends a thought to the universe that is so difficult to comprehend. Neither unit is very complex; they only coexist in their local planetary systems, where things have begun to move and where unprepared units are suddenly faced with the possibility of an expansion, both locally and outside themselves. Here under the night sky everything that seems so difficult to formulate, all subtle features of cosmic communion, becomes quite simple. It is a communion with everything that is struggling and living, wherever and whatever it may be within the frame of the incomprehensibly great. It is not only an incitement to develop but to feel what is already a current concept within the respective worlds, so far from each other.

Compassion. Perhaps.

The darkness of the night links the two units together, separated as they are by millions of light-years. A moment of understanding travels through space with the speed of an expanding thought, relayed by those who have gained a certain insight and sensitivity. A signal conveying a message.

COMMENT

The attempt that has been made—within a rather limited frame—to suggest some tendencies in man's present situation, a conception of the developmental possibilities of our planetary home district, and a view of the concept of life should, of course, be regarded as an experiment. The presentation of astronomical and biochemical data—the unconstrained mixing of which may have confused more than one reader—has one direct objective, that is, to indicate a new direction for our scientific thinking. Certain biochemical problems can no longer be treated from the local point of view but must be seen in a wider perspective. After having been a somewhat dubious subject of discussion, the development of life as a general phenomenon within the universe is now being examined from the point of view of the experts; astrochemistry and astrobiology are gaining ground as specialized professions in several places on our earth. The following books are recommended for a more detailed introduction to the subject.

BROWN, R. H., and LOWELL, A. C. B. *The Exploration of Space by Radio.* London: Chapman & Hall, Ltd., 1957.

CAMERON, A. G. W. *Interstellar Communication.* New York: W. A. Benjamin, Inc., 1963.

DOLE, S. H. *Habitable Planets for Man.* New York: Blaisdell Publishing Co., 1964.

FIRSOFF, V. A. *Strange World of the Moon.* London: Hutchinson & Co., Ltd., 1959.

———. *Surface of the Moon.* London: Hutchinson & Co., Ltd., 1961.

———. *Life beyond the Earth.* London: Hutchinson & Co., Ltd., 1963.

Harvard Books on Astronomy, The. New York: Blakiston Co., 1941–56.

HOYLE, F. *Frontiers in Astronomy.* London: Heinemann Publishers, Ltd., 1955.

HYNEK, J. A. *Astrophysics.* New York: McGraw-Hill Book Co., 1951.

MOORE, P. *Guide to the Moon.* London: Eyre & Spottiswoode, Ltd., 1953.

———. *The Planet Venus.* New York: Macmillan Co., 1959.

PEEK, B. M. *The Planet Jupiter.* London: Faber & Faber, Ltd., 1958.

SMITH, F. G. *Radio Astronomy.* Baltimore, Md.: Penguin Books, Inc., 1960.

STRUGHOLD, H. *The Red and the Green Planet.* London: Sidgwick & Jackson, Ltd., 1954.

UREY, H. *The Planets.* New Haven, Conn.: Yale University Press, 1952.

WIECH, R. E., JR., and STRAUSS, R. T. *Fundamentals of Rocket Propulsion.* ("Reinhold Space Technology Series.") New York: Reinhold Publishing Co., 1960.

Some references to literature dealing with geological and biochemical problems of the origin of life will be found in the reference list of a short book by the author, *Life: Origin and Development,* in which some lines of thought are presented that have been developed further in the present treatise at a different level.

The pervading theme of the present work is the general, universal tendency that can be traced behind the development of conscious life in all its forms. Certain features in this developmental tendency have been accentuated against a clearly realistic background. That an indication of a hopeful prognosis as far as man is concerned can be traced once in a while—especially in the more subjectively colored parts of the presentation—is, of course, a matter of debate. Such elements of unspecific optimism should be taken as irrelevant expressions of the personal views of the author when confronted with an abstruse development with no well-defined goal—an attitude toward the nearest future that is free from illusions but that includes a certain hopefulness with regard to the long-term aim.

INDEX

Albedo value
of earth, 32
of Mercury, 107
of moon, 32, 72, 107
of Venus, 71–72
Amino acid, 21, 22, 56–59, Fig. 7 (58–59)
see also Biochemistry; Protein
Ammonia, 57–60, Fig. 7 (58–59), 107, 113
Aquilae, α-, 130 (Table 16)
Argon-40
on Mars, 86–87, Table 11 (87)
on moon, 34–35, 37 (Table 4)
on Venus, 71 (Table 8)
Asteroids
atmosphere and ground conditions of, 13 (Table 1B)
distance of, from sun (relative), 14 (Fig. 2)
Astronomic unit, 14
Atmosphere
general compositions of, 13 (Table 1B), 19–20
of Jupiter, 107
of Mars, 12, 85–87, Table 11 (87), 89
water loss from, 116–18
of Mercury, 12, 107–9
methods of observing, 11 (Fig. 1), 12 (Table 1A), 18–21
of moon, 13, 34–38, Table 4 (37)
of Titan, 113
of Venus, 65–72, Tables 7, 8 (68, 71), 76–81
Atomic energy; *see* Energy; Nuclear fission
ATP (adenosine triphosphate), 142 (Fig. 9), 143
AU (atmospheric unit), 14 (Fig. 2)
Awareness; *see* Conscious life

Bergmann, T., 66
Binary system (interplanetary communication, 131
Biochemistry, v–vi, 43–44, 47–64, 116–19, 138–39, Fig. 8 (139), 144
and conscious life, universal possibilities of evolution of, 137–53
evolution of, on Venus, 76–81
see also Amino acid; Protein
Bobrownikoff, Nicholas T., 68
Bradbury, Ray, 93
Braking mechanism
of manned spaceship, 30, 89
of stored equipment, 39

"Canals" of Mars, 97
Canis Majoris and Canis Minoris, 130 (Table 16)
Carbohydrates, 21, 52

Carbon, distribution of, 115–17, Table 14 (117)
Carbon compounds
basis of life, 138, 170–71, 115–19
characterization of, 21
distribution of, 68 (Table 7)
effect of temperature on, 54–55
on Mars, 98–101
reaction possibilities, 56
role of, in evolution, 76–81, 138
on Venus, 68, 69 (Table 7), 110–12
Carbon dioxide
basic organic component, 48, 49
in cell, liberation of, 140, 142–43 (Fig. 9)
on Mars, 85, 87 (Table 11), 90, 118
on moon, 35–37, Table 4 (37), 43–44
oxygen production, 26–27, 35, 75, 92–93
production and storage of, 111
result of enzymatic breakdown, 138
on Venus, 67–71, Tables 6, 7, 8 (68, 71), 75, 118
Cell, 52
as function, 141–46
as structure, 140–46, Fig. 9 (142)
Centauri, α-, 127, 130 (Table 16)
Chromatography, 22
Chromosomes, 142 (Fig. 9)
Clouds and cloud cover, of Venus, 28–29, 71–73
see also Atmosphere
Colonization and observation
construction and design of bases, 38–39, 42 (Fig. 4), 75
general, 7–9, 104–14, 156–57
landing possibilities (relative) of planets, 106 (Table 13)
of Mars, 83–103, 104–6
of moon, 38–46, 104
preparations for, 82
problems of, 24–29, 44
storage of material for, 39–40
of Venus, 73–83, 105, Table 13 (106)
Communication, 17
between cells, 143
between earth and outposts, 113–14
among Martians (Stapledon's theory), 147–48
on and from moon, 37–38, 104
of scientific information; *see* Telemetry
between solar systems, 123–36, Table 15 (128)
time required, 114, 132–35
type of information relayed, 131–35
use of computers in, 18, 130–31
Conscious life, v–vi, 8–9, 137–53, 167–72
communication with, 123–53
emigration of, from home planets, 156–57
evolution of, 76–81, 115–19, 138–53, 157–61
human, basis of, 146
on Mars or Venus, 15, 47, 116–19
see also Communication
Cyan compounds, 76–81

Data transmission; *see* Telemetry
Deimos (satellite of Mars), 88–89
Distances, within planetary systems, 12 (Table 1A), 14 (Fig. 2)
DNA (deoxyribonucleic acid), and protein synthesis, 141
Doppler effect, 12 (Table 1A)

Echo (American communications satellite), 38
Electromagnetic radiation, 21 (Fig. 1)
Energy, 50–51 (Fig. 5)
from atomic power plants, 4, 41, 75, 110
conservation of, 24–28
cycles, 49–51, Fig. 5 (50–51)
light, 27–29, 40–41, 43, 51–53, 60–61
solar batteries, 16, 17
sources and conversion of, 15, 16–17, 59–61, 74–75, 111–12, 161
for space flight, 24
Enzymes, 56, 138, 142 (Fig. 9)
Eridani, ε-, Eridani A, γ^2-, Eridani B, γ^2-, and Eridani C, γ^2-, 130 (Table 16), 162–63

Erosion, on moon, 32–33
Evolution; *see* Life; Evolution
Expansion, defined, v–vi

Firsoff, V. A., 33–34
Fission; *see* Nuclear fission
Food, storage of, 25–29, 39–40

Galaxies, 124, 125, 165
Gas chromatography, 20–21
  see also Telemetry
Geologic and topographic conditions; *see* Ground surface conditions
Giant planets, 53, 105–6
Goldschmidt, V. M., 68
Gravitational force, 106 (Table 13)
  effect of, on landing possibilities, 106 (Table 13), 107
  on Mars, 84–85, 91–92
    favors water loss, 116
  on moon, 32 (Table 3), 33
Green Bank, United States, observatory at, 130
Ground surface conditions
  comparative data for giant and smaller planets, 106 (Table 13)
  of J. IV, 110
  of Mercury, 108
  of moon, 31–36, Table 3 (32), 108
  observations of, 11–15, Tables 1A, 1B (12–13), 18–19
Guided missiles, 13

Helium
  in atmospheres of Neptune and Pluto, 113
  solar energy cycle, 156
Hoyle, Fred, 148–50
Huxley, Aldous, 160–61
Hydrocarbons, 57
  basis of life, 115–16
  distribution of, 69, 70
  on J. IV, 110–12
  on Mars, 98–101
  on Venus, 69, 76–81
Hydrogen, 50 (Fig. 5), 51, 53
  in atmospheres of Neptune and Pluto, 113
  in cells, 140
  transformation of, in solar energy cycle, 156
Hydrogen sulfide, energy source for sulfur bacteria, 48–49, 50 (Fig. 5)

Intergalactic communication; *see* Communication
Intergalactic distances
  and communication, 125
  measured in light-years, 123
Iron ore, 111
Iron oxides, 111
  in red areas of Mars, 116, 118

J. III and J. IV (satellites of Jupiter)
  atmosphere of, 110–12
  communication from, 113–14
  landing possibilities and physical characteristics of, 106 (Table 13)
  observation of Jupiter from, 112–14
  orbit distance of J. IV from Jupiter, 109–10
Japetus; *see* S. VII
Jodrell Bank Observatory (England), 10, 127
Jupiter
  atmosphere of, 13 (Table 1B), 70, 107
  combination of elements on, 98
  communication from, 113–14
  compared with earth, 106 (Table 13)
  distance of, from sun (relative), 14 (Fig. 2)
  ground pressure on, 105–7
  landing possibilities of, 106 (Table 13)
  observation of, from J. III and J. IV, 12–13
  rotation period of, 13 (Table 1B)
  satellites of, 110–13
  surface conditions of, 12 (Table 1B), 106 (Table 13)
  temperature of, 13 (Table 1B), 53, 55 (Fig. 6*b*)

Landing
in another planetary system, 135–36
on Mars, 89
on moon, 30–31
possibilities of, on giant and smaller planets, 106 (Table 13)
on Venus, 73–74
Lead-208, 52
Lederberg, J. H., 21–22
Lichens, synthesizing, for growth on Mars, 94–96
Life
general, v–vi, 47–64, 118–19
evolution of, v–vi, 8, 76–81, 137–53
as expression of chemical reaction, 48–64, 115–19
on Mars, 15, 93–101, Table 12 (100–101), 116–19
observation of, 21–27
organization of, 41, 42 (Fig. 4)
on planetary systems, 125–26
survival temperature of, 54 (Fig. 6*a*)
on Venus, 15, 28–29, 55 (Fig. 6*b*), 76–81, 116–18, Table 14 (117)
see also Conscious life
Light energy
artificial, 28
on J. III and J. IV, 111–12
and organic matter, 48, 60–61
power plants for, 41, 43
production of carbon dioxide and oxygen, 27–28
Light ring, around Venus, 66–67
Light-year, 124
Lowell, P., 96–97
Lunik III (Russian unmanned data collector [1959]), 15–16

Manned spaceships, 4
construction of, 15
landing maneuvers of, 30–31, 89, 135–36
problems concerning, 27–30
see also Colonization; Satellites; Space travel; Unmanned data collectors

Mars
atmosphere of, 13 (Table 1B), 84–87, Table 11 (87), 91–93, 116–18
"canals" of, 97
colonization of, 83–103, 104–5, 106 (Table 13)
compared with Venus, 65, 84–85
distance of, from sun (relative), 14 (Fig. 2)
gravitational force of, 84–85, 86, 91–92, 116–18
iron oxides on, 116–18
life on, 15, 47, 55 (Fig. 6*b*), 61–64, 91–101, Table 12 (100–101), 116–19, Table 14 (117)
observation of, 18–30
polar regions of, 91, 98
red areas of, 88, 116, 118
satellites of, 88–89
summary of data about, 13 (Table 1B), 85 (Table 10)
temperature of, 13 (Table 1B), 55 (Fig. 6*b*), 88, 89, 94, 117 (Table 14)
water on, 85–87, 116–18
winds on, 91–93
Mass, 11, 12 (Table 1A)
comparative data for giant and smaller planets, 106 (Table 13)
of Mars, 85 (Table 1)
effect on evolution of, 116–18
of Venus, 66 (Table 5), 73, 75–76
Mass number, used to compute atmospheric components, 20, 22–24, Table 2 (22–23)
Mass spectrometer, 20, 97
MDT (milieu data transmitter); *see* Telemetry
Mercury (element), 109
Mercury (planet)
atmosphere of, 13 (Table 1B), 108
colonization of, 107–9
distance of, from sun (relative), 14 (Fig. 2)
possibilities of, for landing, 106 (Table 13), 107–8
radiation on, 108–9
revolution of, 108

rotation period and surface conditions of, 13 (Table 1B), 108–9
temperature of, 13 (Table 1B), 55 (Fig. 6*b*), 106 (Table 13), 107–9
Metal ions, energy source, 49, 51 (Fig. 5)
Meteorites, possible cause of moon's craters, 15, 33–34
Methane, 57, 58 (Fig. 7), 59, 107
on J. IV, 110–12
on Neptune, 113
on Titan, 113
on Venus, 69
Microsomes, 140–43, Fig. 9 (142)
Milieu data transmitter; *see* Telemetry
Milky Way, 124–25
Miller, S. L. I., 57
Mineral salts, 48, 52
Mirrors, 10, 11 (Fig. 1), 12 (Table 1A)
Mitochondria, 140–43, Fig. 9 (142)
Moon (earth's)
albedo value of, 32, 71
atmosphere of, 13 (Table 1B), 34–38, Table 4 (37)
colonization of, 38–46, 104
communication on, 37–38
craters of, 15, 33–34
distance of, from sun (relative), 14 (Fig. 2)
erosion on, 32–33
gravitational force on, 32–33
hidden surface of, 16
landing on, 31–32
orbit of, 16, 32 (Table 3)
photographs of, 16–17
summary of data about, 32 (Table 3)
surface conditions of, 13 (Table 1B), 31–36, Table 3 (32)
temperature of, 13 (Table 1B), 32, 38, 43–44
water on, 25
Moons; *see* J. III and J. IV; S. VII
Mount Palomar Observatory, 10, 18, 98, 127–28
Mount Wilson Observatory, 67
Mountains, on moon, 31–34

Neptune
atmosphere of, 13 (Table 1B), 113
distance of, from sun (relative), 14 (Fig. 2)
landing possibilities of, 106 (Table 13)
satellite of, 113
summary of data about, 13 (Table 1B)
temperature of, 13 (Table 1B), 55 (Fig. 6*b*), 106 (Table 13)
Nitrogen
compounds of, 138
on Mars, 86–88, Table 11 (87)
on moon, 37 (Table 4)
on Venus, 70, 71 (Table 8), 76–81
Nobelium-252, 52
Noble gases, on Mercury, 109
Nuclear fission, 3, 4
power plants using, 4, 28, 41, 42 (Fig. 4), 111–12
see also Energy
Nucleic acid, 52

Observation methods, 10–29
see also Telemetry
Orbits, 11, 12 (Table 1A)
radius of moon's, 32 (Table 3)
radius of Mars's, 85 (Table 10)
radius of Venus', 66 (Table 5)
Oxygen
in cell, 52, 140, 142 (Fig. 9)
on Mars, 85–87, Table 11 (85)
production and storage of, 26–27, 39–40, 48–49, 90–92, 111, 116–18
role in evolution of, 138

Parabolic reflector transmitter; *see* Telemetry
Perception and consciousness, 137–53
see also Conscious life
Phobos (satellite of Mars), 88–89
Photographs
data from, 16, 18–21
of moon, 16
use of, to map Mars's surface, 97

Photolysis, of water, 60–61
on Mars and Venus, 116–18
Photosynthesis, 48–52, Fig. 5 (50–51), 159, 161, 163, 170–71
on Mars, 118
on moon, 40–44
on Venus, 77, 118
Phototelevision techniques; *see* Telemetry
Planetary system, 125–26
distances within earth's, 14 (Fig. 2)
Pluto
atmosphere of, 13 (Table 1B), 113
communication from, 114
compared with earth, 113
distance of, from sun (relative), 14 (Fig. 2)
landing possibilities of, 106 (Table 13)
Lowell's calculations about, 97
mass of, 113
rotation period and surface conditions of, 13 (Table 1B)
temperature of, 13 (Table 1B), 55 (Fig. 6*b*), 113
use of, as base, 113
Plutonium, 52
Polonium-210, for electricity production, 217
see also Energy; Nuclear fission
Potassium-40, source of argon-40
on Mars, 86–87
on moon, 34–35
on Venus, 71
Power supply; *see* Energy; Nuclear fission
Probes; *see* Unmanned data collectors
Protein, 52, 56, 57, 58 (Fig. 7), 59

Radar, 12 (Table 1A), 73
Radiation
danger of, from nuclear power plants, 41
on Mars, 109
range of, within which life can exist, 125–26
relative intensity of, within planetary systems, 130 (Table 16)
Rankama, K., 68
Reflection capacity; *see* Albedo value
Reflectors, 10, 11 (Fig. 1), 127–29
on moon, 45
Research, 3–9, 15
Ring formations, on moon, 33–34
Robot satellites; *see* Satellites
Rockets, 3, 15, 19–21
see also Satellites; Space travel; Unmanned data collectors
Rotation elements, 11, 12 (Table 1A), 13 (Table 1B)

S. VII (Japetus, satellite of Saturn), 113
Satellites, 4, 15–19, 25 (Fig. 3)
Echo, 37–38
Lunik III, 15–16
robot, 15
see also J. III and J. IV; S. VII; Telemetry; Titan; Unmanned data collectors
Saturn
atmosphere of, 13 (Table 1B), 70
combination of elements on, 98
distance of, from sun (relative), 14 (Fig. 2), 106 (Table 13)
landing possibilities of, 106 (Table 13)
physical characteristics of, 106 (Table 13)
rotation period and surface conditions of, 13 (Table 1B)
satellite of, 113
temperature of, 13 (Table 1B), 55 (Fig. 6*b*), 106 (Table 13)
Schiaparelli, G., 96–97
Shapley, H., 35
Silicium, 52
Sinton, W. M., 98
SNAP (Systems for Nuclear Auxiliary Power) project, 16–17
Social systems of future, 93–94, 146–50, 154–64
Solar batteries, 16, 17
Solar energy; *see* Light energy
Sounding projectiles, 30
Space colonization; *see* Colonization and observation

Space probes; *see* Unmanned data collectors
Space travel, 123
  current developments, 15
  landing maneuvers, 30–31, 135–36
  for observation, 126
  problems of survival during, 24–29
  see also Manned spaceships; Unmanned data collectors
Spaceships; *see* Manned spaceships; Space travel
Spectral class (relative) of planetary systems, 130 (Table 16)
Spectrometer; *see* Mass spectrometer
Spectroscope, 67
Spectrum analysis, 10, 11 (Fig. 1), 12 (Table 1A), 127
Stapledon, Olaf, 93–94, 146–50, 160
Star clusters and star systems, 124–25
Stratospheric balloons, 71
Strontium-90, 17
Strughold, H., 94–96
Sulfur, 53
Sulfur dioxide, on moon, 37 (Table 4)
Sun
  energy from, 41, 42 (Fig. 4), 43–44, 51–53, 155–66
  obtaining data about, from Mercury, 109
  period of moon's revolution about, 32 (Table 3)
  and photosynthesis, 48–52
  as seen from Pluto, 113
  temperature of surface of, 54 (Fig. 6*a*)
Surface conditions of planets; *see* Atmosphere; Ground surface conditions; Temperature
Synthesis mechanisms of cells, generalized, 49, 50 (Fig. 5)
Systems for Nuclear Auxiliary Power project; *see* SNAP

Technocratic society, speculation about, 154–55
Telemetry, 16–21, 30, 31, 126, 134
  gas chromatography equipment, 20–21
  MDT, 22 (Table 2), 23–24
  parabolic reflector transmitter, 17
  of photographic data, 16
  phototelevision techniques, 18
  see also Communication
Telescopes, 11 (Fig. 1), 12 (Table 1A), 70, 127–29, 132
Television, 18, 21, 31
Temperature
  atmospheric, 13 (Table 1B), 55 (Fig. 6*b*)
  comparative data for giant and smaller planets, 105, 106 (Table 13)
  and development of planet, 53–55 (Figs. 6*a*, 6*b*), 63
  of ground surfaces, 12 (Tables 1A, 1B)
  on Mars, 89, 94
  on Mercury, 107–8, 109
  on moon, 32 (Table 3)
  in Neptune-Pluto region, 113
  on S. VII, 113
  survival, 53–55
  on Venus, 72–73, Table 9 (73)
Thought; *see* Conscious life
Titan (satellite of Saturn)
  atmosphere of, 113
  bases on, 113
  landing possibilities and physical characteristics of, 106 (Table 13)
Topographic and geologic conditions; *see* Ground surface conditions
Transmitters; *see* Telemetry
Triton (satellite of Neptune), 106 (Table 13), 113

Unmanned data collectors, 4, 10–15, 24, 25 (Fig. 3), 28, 65
  Lunik III, 15–16
  to Mars, 88
  to other galaxies, 126
  SNAP project, 16–17
  stratospheric balloons, 71
  Venus fly-by mission, 4, 18
  see also Manned spaceships; Satellites; Space travel; Telemetry
Uppsala observatory (Sweden), 66
Uranium-238, 52, 60

Uranium fission; *see* Nuclear fission
Uranus
atmosphere of, 13 (Table 1B)
distance of, from sun (relative), 14 (Fig. 2), 106 (Table 13)
landing possibilities and physical characteristics of, 106 (Table 13)
rotation period and surface conditions of, 13 (Table 1B)
temperature of, 13 (Table 1B), 55 (Fig. 6*b*), 106 (Table 13)
Urey, H., 33

V-2 (German World War II rocket), 3
Vegetation, 48–52
cultivation of, 11–12, 75
and oxygen production, 28, 91–92
survival of, on Mars, 94–101
see also Life; Photosynthesis
Venus
albedo value of, 71–72
atmosphere of, 13 (Table 1B), 28–29, 65–72, 71 (Table 8), 76–81, 104, 116–19
atmospheric and ground pressure of, 67–68, Table 6 (68)
carbon material on, 68 (Table 7)
cloud cover, 28–29, 71–73
colonization of, 73–83, 105, 106 (Table 13)
compared with earth, 66 (Table 5)
compared with Mars, 65, 84–85
distance of, from sun (relative), 14 (Fig. 2)
gravitational force of, 66
life on, 15, 28–29, 47, 55 (Fig. 6*b*), 61–64, 76–81, 116–19, Table 14 (117)
light ring around, 66–67
obtaining data about, 18–23, 24–30, 65
rotation period and surface conditions of, 13 (Table 1B)
summary of data about, 66 (Table 5)
temperature of, 13 (Table 1B), 28–29, 55 (Fig. 6*b*), 66, 69–70, 71–73, Tables 8, 9 (71, 73), 105, 117–18
water on, 70–73, Table 8 (71)
Venus fly-by mission (1963–64), 4
see also Satellites; Telemetry; Unmanned data collectors
Volcanoes and volcanic activity, on moon, 33–34, 35–36

Water
basic organic component, 48
in cell, 138, 140–43, Fig. 9 (142)
distribution of, 116, 117 (Table 14)
on Mars, 85–87, 91–93, 116–18
on moon, 35–37, Table 4 (37), 43–44
and oxygen production, 26
photolysis of, 60–61, 116–18
production and storage of, 26–28, 111
on Venus, 60–61, 70–72, Table 8 (71), 76–81, 116–18
Weapons, 3
Weather; *see* Ground surface conditions; Temperature; Wind
Wells, H. G., 93
Wind
on Mars, 91–93
on moon, 32
on Venus, 73
see also Ground surface conditions; Temperature

Xenon, 109